AI NEW WORLD

Expanding the Limits of Knowledge

First published in English in 2023 By Mekiki Magazine

Site: www.mekikimagazine.com

ISBN: 979-8-89214-009-6

Title: AI New World: Expanding the Limits of Knowledge
Author: Mekiki
Editor: A. Lee

Book Cover By: Álvaro Oliveira For Mekiki Magazine
Graphic Design: Álvaro Oliveira For Mekiki Magazine

As the author, we want to disclose that we employ AI tools to aid in editing and improving our writing. These AI tools, including editing, grammar, and spell-checking, assist us in enhancing the quality and readability of our work. However, it is essential to acknowledge that automated systems are not infallible. While we strive for accuracy and clarity in our writing, it is advisable for readers to exercise their own judgment and critical thinking. Our commitment remains focused on providing the best possible reading experience, and we sincerely appreciate your understanding and support in this endeavor.

CONTENTS

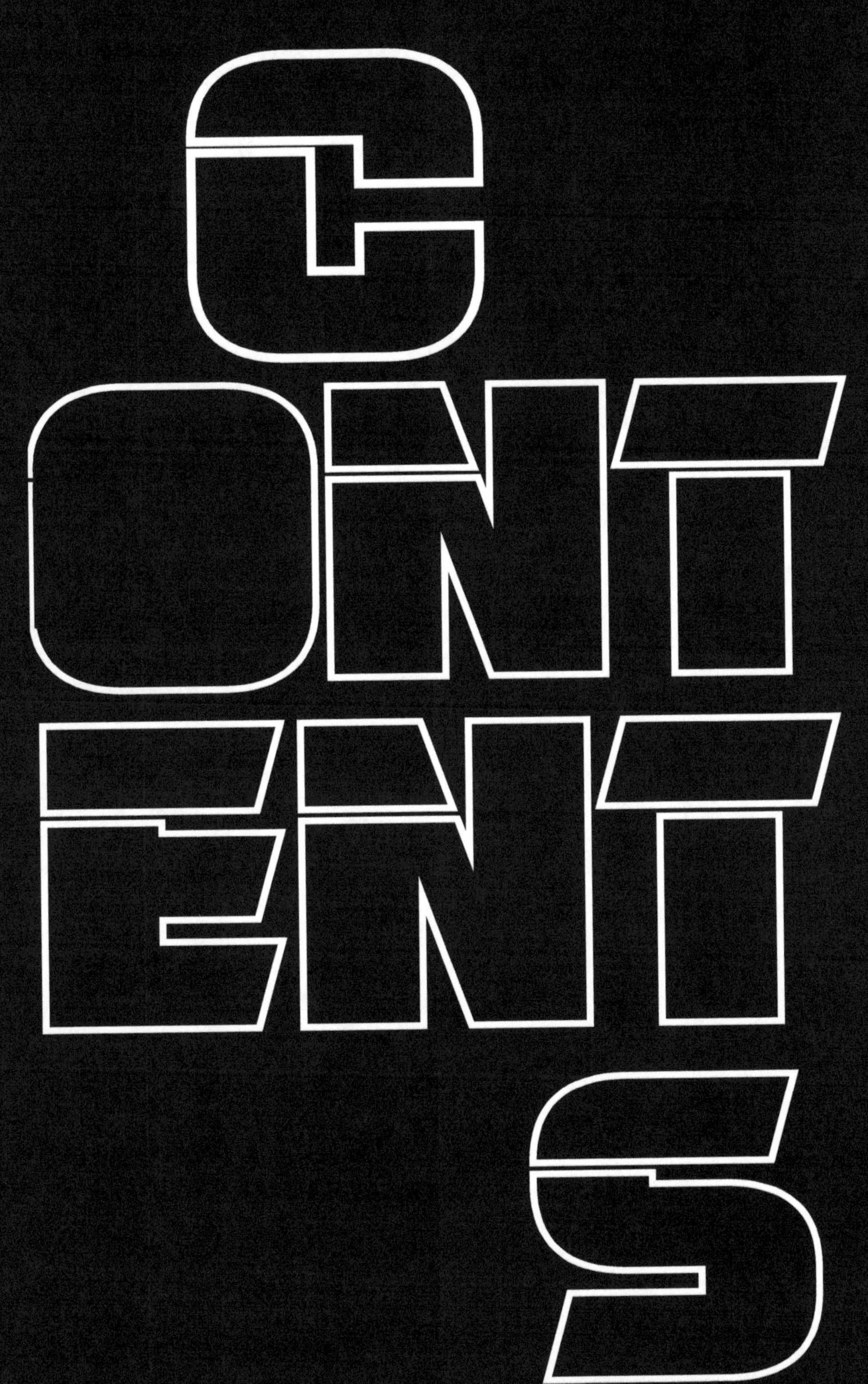
CONTENTS

ARTIFICIAL

INTRODUCTION

INTELLIGENCE

As I pen these words, the world outside my window carries on as it always has. Birds are singing their morning serenades, cars hum in the distance, and the sun embarks on its westward journey across the sky. All is as it should be, and yet, beneath this veneer of normalcy, an unseen revolution is taking place—a revolution set to change the very fabric of our existence. Welcome, dear reader, to the dawn of the AI era.

Artificial Intelligence, or AI, as it is fondly known, is no longer a distant dream confined to the realm of science fiction. It is here, living amongst us, subtly weaving its magic into the tapestry of our daily lives. From the moment you ask your virtual assistant for the day's weather forecast to the second you unlock your smartphone with a glance, you are interacting with AI. Unassuming yet powerful, AI is steadily transforming the way we live, work, and think.

Yet the story of AI is far from linear. It is a tale steeped in paradoxes, a narrative that is as exhilarating as it is daunting. AI offers limitless opportunities, such as cities that are in sync with us, education that adapts to each learner's pace, and health care that can predict ailments before they appear. On the other hand, it forces us to grapple with profound questions about job displacement, privacy concerns, and ethical dilemmas.

This book is an exploration of that story and a thoroug
trip that will take us from the buzzing tech hubs of Silic
of the Global South. From the hallowed halls of acade
labs, pushing the boundaries of what is possible. Alon
disruptors, dreamers, and dissenters who are shaping

However, this book is not just about AI; it is about us—h
this brave new world with wisdom, empathy, and foresig
that will determine the kind of tomorrow we bequeath to

Whether you are a seasoned technolo
sense of this rapidly evolving landscap
engage with the most transformative te

So buckle up and brace yourself for an
and embrace it. As you turn this page, y
human history. Welcome aboard, dear r

lysis of the intriguing world of AI It is a
lley to the vibrant startup ecosystems
the most up-to-date research
vay, we will meet the innovators,
landscape.

nity. It reflects how we can navigate
s about the choices we make today
generations.

a curious observer, a policymaker, or a student trying to make
is book is for you. It is an invitation to dialogue and a call to
ology of our time.

ing adventure. The AI revolution is here, and it is time to lean in
re not just starting a new chapter in a book but in the annals of
r. Our exploration of the future begins now.

WHY THIS BOOK?

The need for a comprehensive, multidimensional perspective on AI

Immersed in the rhythmic clatter of keys that echoes my thoughts onto the digital canvas of our era, I find myself contemplating a question that you may very well be asking yourself—why this book? In a world brimming with an almost infinite sea of information, where the discourse on artificial intelligence (AI) ebbs and flows through academic journals, news articles, podcasts, and tweets, why devote precious hours to this particular narrative?

The answer, I believe, lies in the paradox that is AI itself. A paradox where the most profoundly transformative technology of our era is often the least understood, its potential the most underestimated, and its risks the most underappreciated. We are living in an age where AI is no longer an abstract and futuristic concept.

It is here, subtly permeating the fabric of our existence, from the smartphones in our pockets to the digital assistants in our homes, from the algorithms that curate our online experiences to the automated systems that underpin our economies.

Yet despite AI's ubiquitous presence, there exists a yawning chasm between the world of AI and the world at large. A chasm born out of inaccessible jargon, a seemingly insurmountable learning curve, and a story that often oscillates between irrational fear and unfounded hype. This book is my humble attempt to bridge that gap—to make the complex world of AI not just accessible but also engaging and relevant to you, the reader.

But this is not just a book about AI. It is a book about us. The story of AI is intrinsically intertwined with our own narrative, reflecting our biases, amplifying our capabilities, and challenging our ethical constructs. It is a narrative that forces us to confront not just who we are but also who we aspire to be.

This book aims to provide a comprehensive, multidimensional perspective on AI. It digs not just into the "how" of AI—algorithms, data, and models—but also the "why"—the societal implications, ethical dilemmas, and policy challenges. It explores AI's potential to both empower and disenfranchise, to create and destroy, to liberate and control.

Furthermore, it draws upon a diverse range of voices—from computer scientists to ethicists, educators to policymakers, industry veterans to startups—to paint a holistic picture of the AI landscape. It navigates the past and the present to illuminate possible futures, examining how AI has developed, where it stands today, and where it could take us tomorrow.

Whether you are a student grappling with the fundamentals, a professional keen on leveraging AI in your industry, a policymaker navigating the complex terrain of AI governance, or a curious mind intrigued by the promise and peril of AI, this book is for you. By reading this book, I hope you will not just learn about AI but also engage with it, question it, and shape it. The story of AI is not just a story about technology. It is a story about people, about society, and about us. And it is a story that is too important to be left to a privileged few.

So I invite you to go on this journey with me, a route that traverses the intriguing landscape of AI, a journey that explores not just the mind of the machine but also the heart of humanity. After all, the narrative of AI is still being written, and we all have a part to play in shaping its plot.

WHO SHOULD READ THIS BOOK?

Standing at the edge of our collective venture into artificial intelligence (AI), I reflect on who this exploration is for. Who should accompany us on this expedition into the very heart of our modern digital society? Who will derive the most benefit from this meticulously woven narrative?

The answer is as complex and multifaceted as AI itself. This book, with its expansive lens and approach, is designed to resonate with a broad spectrum of individuals, each bringing their unique perspective to the shared discourse on AI.

If you are a student, whether in high school or university, grappling with the increasingly important role that AI plays in our world, this book is for you.

It will lead you through the complex and intricate maze of AI, making complex concepts understandable and interesting. This book will provide you with the foundational knowledge you need to explore deeper in this fascinating field, sparking your curiosity and fostering a lifelong love of learning.

For professionals, whether you are a seasoned software engineer, a marketing executive, a health care professional, or an entrepreneur, this book is your companion. It offers insight into how AI is transforming your industry and provides guidance on leveraging AI's potential to innovate, disrupt, and excel. In these pages, you will find not only technical knowledge but also discussions on the ethical, legal, and societal implications of AI in professional contexts.

Policymakers, this book is your ally. In the rush to legislate and regulate AI, it is crucial to understand the technology, its capabilities, and its limitations. This book offers a balanced perspective, presenting the potential benefits of AI alongside the potential risks and challenges. It presents a way forward for a policy that is both informed and effective, balancing innovation with protection and progress with ethics.

For the philosophers, ethicists, and sociologists among you, this book is a treasure trove. It searches for the profound questions that AI poses about consciousness, bias, morality, and society. It invites you to engage with the philosophical debates that are increasingly relevant to our AI-infused world.

Finally, if you are a layperson curious about the buzzword "AI" that seems to permeate every aspect of modern life, this book is your guide. It will take you on a journey from the basics of AI to its potential future, demystifying jargon and grounding abstract concepts in real-world examples.

This book is for anyone who is a part of our modern, interconnected world. AI is not an isolated field applicable only to computer scientists or tech enthusiasts. It is a pivotal part of our societal fabric, influencing everything from health care to education, economics to ethics, and policy to privacy.

Whether you are an expert in the field, an interested onlooker, or anyone in between, this book has something for you. Because in the grand narrative of AI, we all have a role to play, and through understanding, we can ensure that the role leads us to a future where AI is used responsibly, ethically, and to the benefit of all.

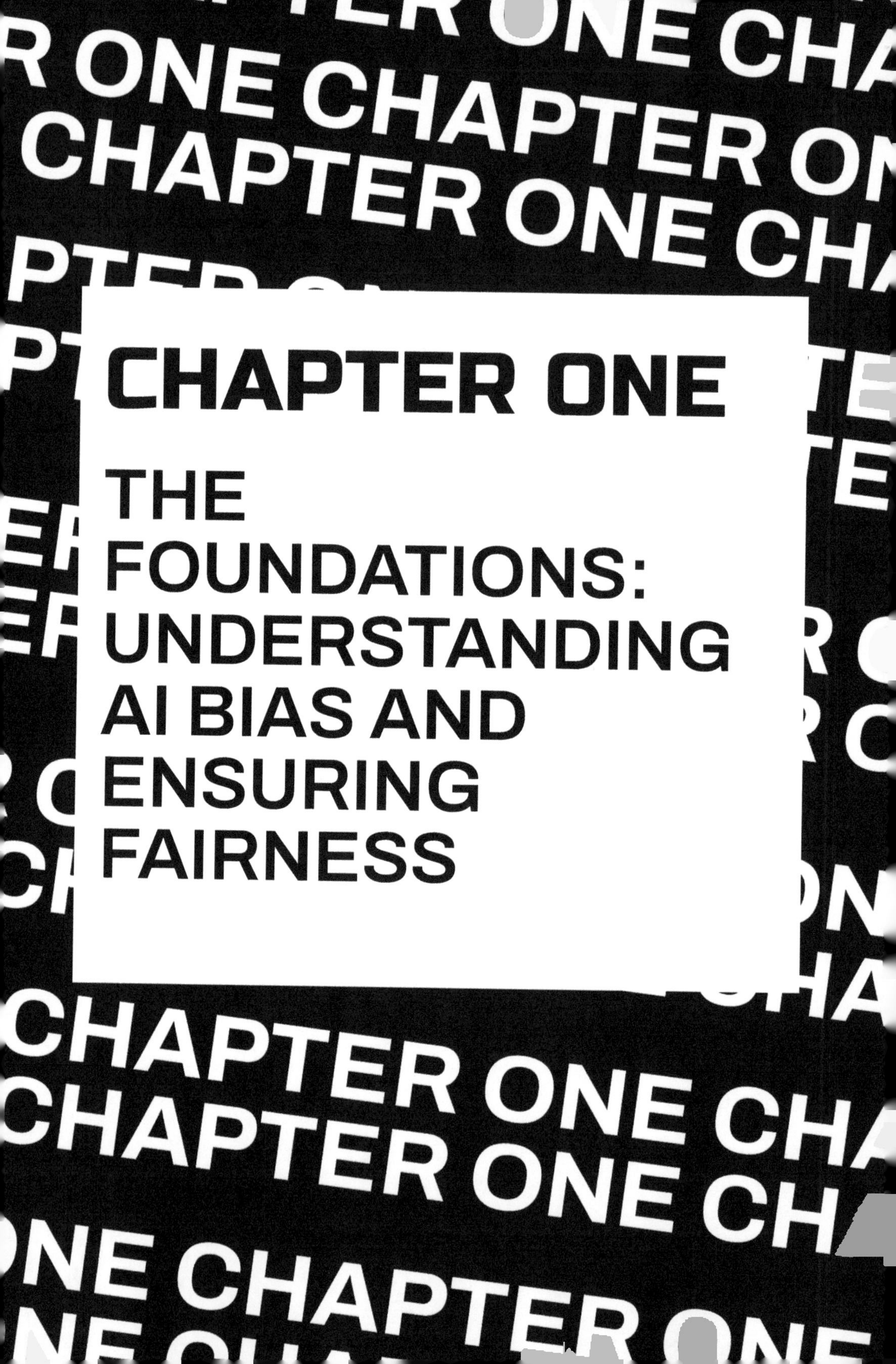

CHAPTER ONE

THE FOUNDATIONS: UNDERSTANDING AI BIAS AND ENSURING FAIRNESS

Unraveling the sources of AI Bias

Before we explore the complex world of artificial intelligence, it is crucial to examine the issue of bias woven into its fabric. Unconscious or not, bias forms the undercurrent of human decision-making, subtly guiding our choices and judgments. In the world of AI, where decision-making is increasingly outsourced to algorithms, understanding and addressing bias is not just crucial. It is imperative.

Bias in AI is not an isolated phenomenon but rather a complex interplay of various factors. To fully comprehend its origins, we need to probe the three primary sources of AI bias: data bias, algorithmic bias, and systemic bias.

Data bias arises from the information we feed into our AI models. AI learns from data. If the data it learns from is skewed, incomplete, or unrepresentative, the AI will reflect and perpetuate these biases. Consider, for instance, facial recognition systems trained primarily on light-skinned male images. Unsurprisingly, these systems struggle to accurately recognize individuals who deviate from this "norm," exhibiting high error rates for dark-skinned and female faces.

Next, we encounter algorithmic bias, which stems from the design and structure of AI algorithms themselves. Even if trained on perfectly balanced data, an algorithm may still display bias if it is designed in a way that favors certain outcomes or characteristics. For example, a hiring algorithm might be unintentionally designed to favor candidates from prestigious universities, thereby perpetuating the bias towards candidates with privileged backgrounds.

Finally, systemic bias refers to the broader social, cultural, and institutional biases that influence AI development and deployment. For instance, an AI model developed in a Western context may not perform well when applied in a non-Western context, reflecting the cultural biases and assumptions of its creators.

Unraveling these sources of AI bias is a formidable task, but it is also an opportunity. By acknowledging and addressing bias, we can strive to create AI systems that are fairer, more equitable, and truly reflective of the diverse world they inhabit. The journey may be complex and the terrain challenging, but the destination—a world where AI serves all of humanity and not just a select few—is worth the effort.

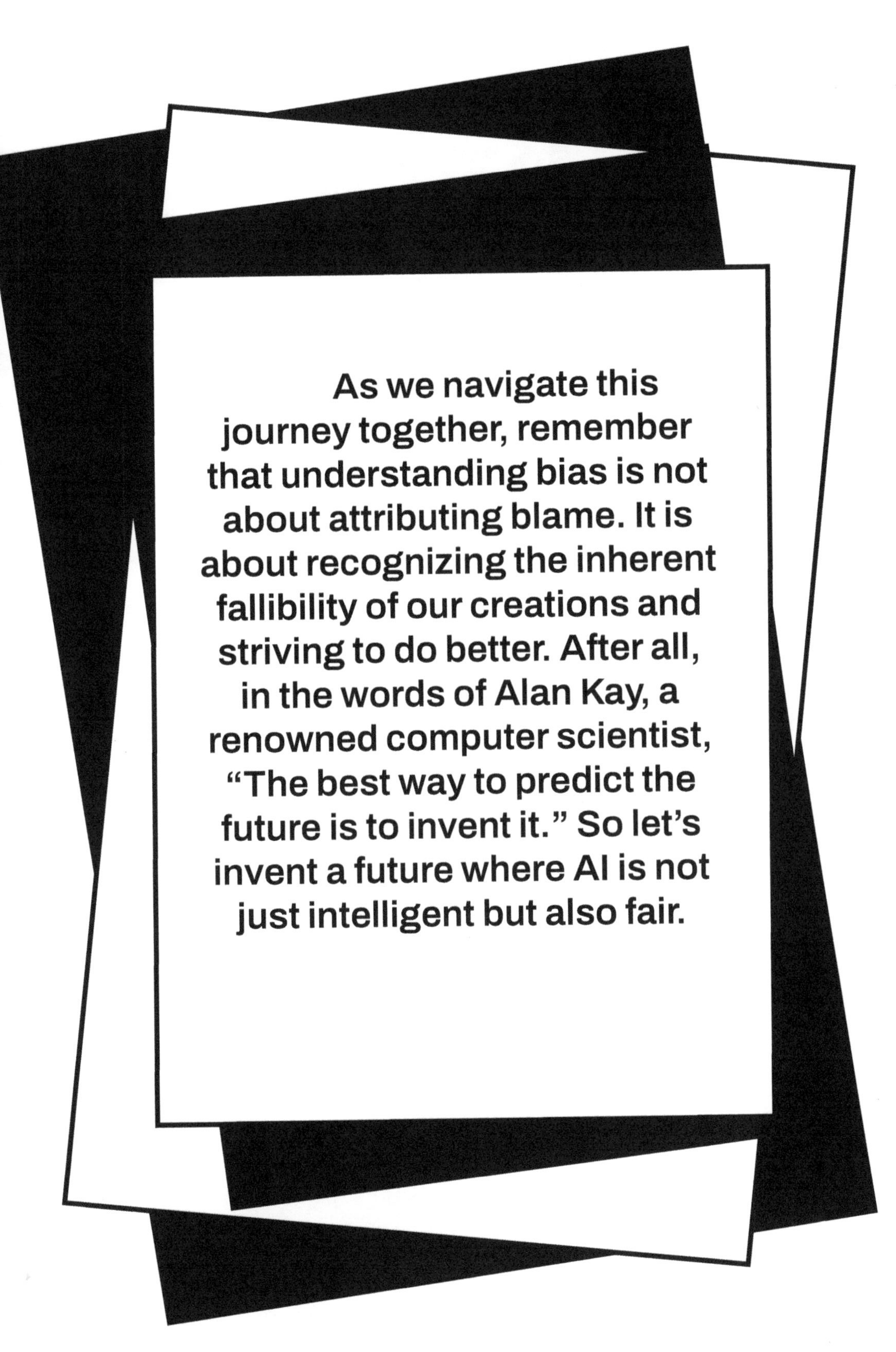

As we navigate this journey together, remember that understanding bias is not about attributing blame. It is about recognizing the inherent fallibility of our creations and striving to do better. After all, in the words of Alan Kay, a renowned computer scientist, "The best way to predict the future is to invent it." So let's invent a future where AI is not just intelligent but also fair.

Real-world impacts of biased AI systems

In the domain of artificial intelligence, bias is not just an academic concern. It is not confined to lines of code or terabytes of data. It transcends the digital world and seeps into our physical reality, casting long, often invisible, shadows on our daily lives. As AI systems become increasingly integral to our societies, the impacts of bias in these systems are felt more profoundly than ever before.

Consider the case of health care, a field where AI holds enormous promise, from predicting disease outbreaks to personalizing treatment plans. But what happens when the AI systems making these decisions are biased? For example, a 2019 study published in the journal Science revealed that an AI system used to guide health care decisions was less likely to recommend additional medical care for Black patients than for White patients with the same health conditions. The AI was not explicitly programmed to consider race. It was biased because it was trained on cost data, which reflected systemic bias in health care access and utilization.

In the domain of criminal justice, AI-powered risk assessment tools are used to predict the likelihood of an individual committing a future crime. These tools can influence decisions about bail, sentencing, and parole. Yet investigations have uncovered racial biases in these tools, with algorithms wrongly classifying Black defendants as "high risk" occurs at nearly twice the rate as White defendants.

Biases in AI also significantly affect the job market. Many companies use AI-powered systems to screen resumes and predict job performance. If these systems are trained on historical hiring data, they can easily learn and perpetuate existing biases, systematically disadvantaging certain groups of people. For example, an AI might learn that the most successful software engineers in its training data are male and, as a result, downgrade female applicants for such roles.

The proliferation of facial recognition technologies presents another arena where bias can have grave consequences. These systems, known to perform less accurately for individuals with darker skin tones or for women, are increasingly used in surveillance and law enforcement, potentially leading to misidentifications and unjust outcomes.

These examples are not outliers. They are symptomatic of a broader issue: the real-world impacts of biased AI systems. They show us that the bias in AI is not just about fairness; it is about justice, equality, and human rights.

The intention of this book is not to evoke fear or distrust in AI but rather to promote awareness and understanding. As we venture further into the world of AI, let us be mindful that we are not only shaping technology but also our collective future. In this future, there should be no room for bias. This book emphasizes the urgent need for awareness, understanding, and action as we navigate the evolving landscape of AI.

Strategies for mitigating bias: a technical and societal approach

Addressing bias in AI is akin to untangling a complex knot; it requires both finesse and a multifaceted approach. Mitigating bias is not just a technical challenge; it is a societal one, demanding strategies that span data collection, model design, implementation, and beyond.

Let us begin with the technical aspects. One of the most direct ways to mitigate bias is at the data level. This includes techniques like oversampling underrepresented groups, undersampling overrepresented groups, or synthetically creating balanced datasets. However, this is not always feasible or ethical, particularly when sensitive attributes like race or gender are involved.

Another approach lies in the design and training of AI models. Algorithms can be tweaked to reduce bias—for instance, by introducing a fairness constraint during training or by post-processing the model's predictions to ensure fair outcomes. In recent years, a branch of AI known as "fairness-aware machine learning" has emerged, focusing specifically on these and other techniques.

Yet, as we have seen, bias in AI is not just a matter of data or algorithms. It is also about the systems and structures within which AI is developed and deployed. Therefore, to effectively mitigate bias, we need to look beyond the technical domain and consider societal strategies.

A key societal strategy is diversity and inclusion in AI development teams. Having a diverse group of people involved in designing and building AI systems can help to surface and challenge unconscious biases and assumptions. It is about ensuring a multitude of voices and perspectives are heard, not just those of a privileged few.

Regulation is another important societal tool. Governments and regulatory bodies can play a critical role in setting guidelines and standards for fair AI as well as auditing AI systems for compliance. However, regulation must strike a delicate balance—it should protect against harmful bias without stifling innovation.

We must promote a culture of transparency and accountability in AI. This includes clear documentation of data sources, methodologies, and decisions, as well as mechanisms for auditing and challenging AI outputs. Open-source software, peer review, third-party audits, and "AI explainability" tools can all contribute to this culture.

In our quest to mitigate bias in AI, we must remember that there is no silver bullet, no one-size-fits-all solution. Each use case is unique, and what works in one context may not work in another. The challenge is complex; the stakes are high, but the rewards are even higher—a world where AI is not just intelligent but also fair and just. As we navigate this journey, let us remember that the technology we create is a reflection of us, its creators. If we strive for fairness in ourselves, we can instill fairness in our creations.

Expert insights and case studies

Expert insights and real-world case studies help us distill abstract concepts into tangible realities. They bring to life the human experiences behind the algorithms and data, reminding us of the why behind our pursuit of fair AI.

Let us begin with the expert insights from Dr. Timnit Gebru, an esteemed researcher known for her groundbreaking work in AI ethics. Dr. Gebru's research has underscored the importance of interdisciplinary teams in AI. She asserts that technologists alone cannot tackle bias, emphasizing the need to involve social scientists, anthropologists, ethicists, and others who understand the complex nuances of human societies. Gebru's work is a powerful reminder that AI has human implications and is essentially a human endeavor.

Now, let us turn to a persuasive case study: Google's AI-powered sentiment analyzer. Initially, this AI system was observed to associate negative sentiment with certain phrases related to gender, race, or religion. Recognizing this issue, Google undertook a comprehensive bias mitigation project. Their approach was threefold: refining the model training process, incorporating diverse human reviewers into system development, and continuously testing for bias. This effort resulted in substantial improvements in the system's fairness as well as valuable lessons about the ongoing nature of bias mitigation.

Another enlightening expert perspective comes from Joy Buolamwini, founder of the Algorithmic Justice League. Buolamwini has highlighted the dangers of "coded gaze," the bias present in facial recognition systems. Her research revealed significant disparities in the performance of these systems across different demographics, triggering important conversations about algorithmic fairness. Buolamwini's work underscores the need for continuous scrutiny of AI systems, even after they are deployed.

A case study that illustrates this point well is the COMPAS risk assessment tool used in the criminal justice system. Despite claims of neutrality, analysis showed the algorithm was biased against African American defendants. The controversy sparked a public debate on the transparency, fairness, and accountability of AI systems, influencing policy discussions and inspiring further research into fair machine learning techniques.

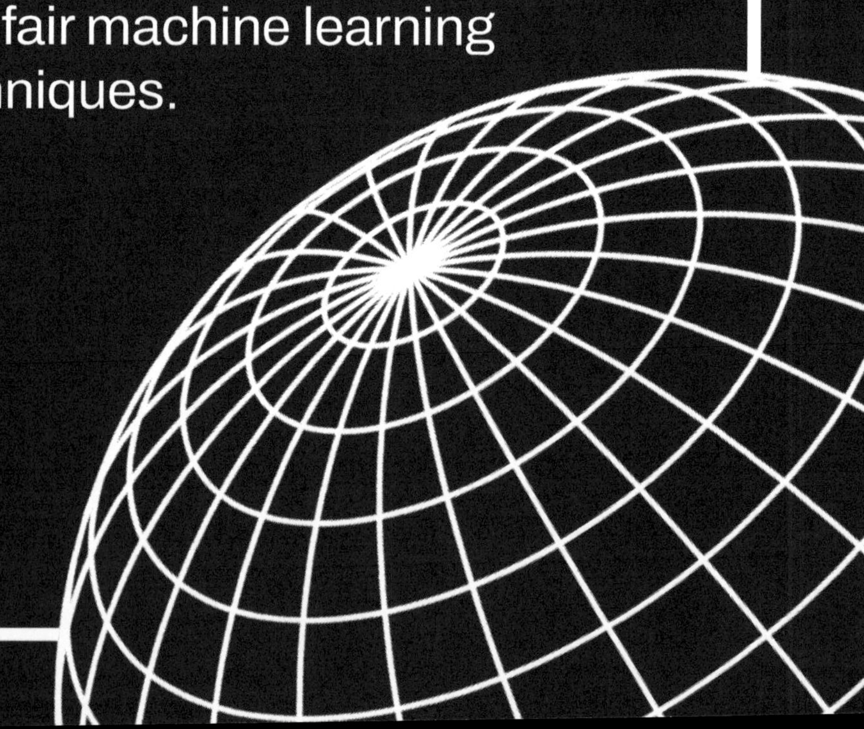

These expert insights and case studies illuminate the path toward fair AI. They remind us that mitigating bias is not just a theoretical exercise but a practical, ongoing commitment. They show us that while the journey may be complex, the destination—a world where AI serves all of humanity fairly—is within our grasp. As we forge ahead, let us draw inspiration from these pioneers and remember that each of us has a role to play in shaping a fair AI future.

CHAPTER TWO

DEMYSTIFYING AI: TOWARD TRANSPARENCY AND EXPLAINABILITY

WHY TRANSPARENCY AND EXPLAINABILITY MATTER IN AI

In the heart of a dense, ancient forest, there is a clear, murmuring stream. The water is so crystalline that you can see the pebbles at the bottom and the fish that dart between them. This stream represents the ideal AI system—transparent and explainable. But why does this matter? Why should we aspire for our AI systems to be like this clear stream?

Transparency and explainability in AI matter for several crucial reasons, both practical and ethical. We will explore these reasons by peeling the layers of this argument like an onion.

The first layer is the most immediate one—trust. AI is increasingly making decisions that have deep impacts on people's lives, from health care diagnoses to job applications to credit scores. If people do not understand how these decisions are made, they are unlikely to trust them. Without trust, the adoption and acceptance of AI systems are hindered, stunting their potential benefits.

The second layer relates to fairness and bias, themes we explored in the previous chapter. Transparency allows us to scrutinize AI systems for bias or unfairness. It enables us to see how decisions are made and whether certain groups are systematically disadvantaged. Without transparency, bias can remain hidden, perpetuating injustice.

The third layer is about accountability. When things go wrong—as they inevitably do—transparency and explainability help us determine where the blame lies. Was it a problem with the data, the algorithm, the implementation, or the user? Without this understanding, it is hard to rectify issues, learn from mistakes, or hold the responsible parties accountable.

The fourth layer is related to learning and improvement. Explainability allows us to understand why an AI system performs well in some areas and poorly in others. This insight can guide further research and development, helping us improve the system and advance the field.

Finally, at the core of the onion is the ethical principle of respect for autonomy. As autonomous beings, we have the right to understand the systems and processes that affect our lives. If a machine denies us a job or a loan, we deserve to know why. By making AI systems transparent and explainable, we uphold this fundamental human right.

Thus, transparency and explainability in AI are not luxuries but necessities. They are the stones that build a bridge between AI and its human users—a bridge of trust, fairness, accountability, learning, and respect. In the upcoming chapter, we will explore the construction of this bridge, methodically building it brick by brick and stone by stone. Let us continue the journey under the guidance of explanation and transparency.

Techniques and tools for improving AI explainability

Imagine being handed a toolbox. Within this toolbox are the instruments you need to make AI systems as clear as the stream we envisioned earlier, allowing the world to see and understand the mechanisms that drive their operation. In this section, we will explore the tools at our disposal and understand the procedures that can help improve the explainability of AI.

The first tool we encounter is known as feature importance. This technique involves identifying which inputs to an AI model have the most influence on its outputs. For example, in a model predicting house prices, feature importance could reveal that the number of bedrooms is more influential than the color of the exterior. Feature importance provides a basic yet significant insight into what an AI system is "paying attention to."

Next, we find a set of tools known as "model-agnostic methods." These are techniques like LIME (local interpretable model-agnostic explanations) and SHAP (SHapley Additive exPlanations), which aim to clarify any AI model by approximating its decisions with simpler, interpretable models. By focusing on local interpretations around specific predictions, these methods can provide insights even into complex black-box models.

Further into the toolbox, we find visualization techniques. These are methods for visually representing the inner workings of AI models. For example, in a convolutional neural network used for image recognition, visualization can help us see what features each layer of the network is learning. Visualization brings the abstract mathematics of AI into the tangible realm, making it more accessible and comprehensible.

Among these tools, we also discover counterfactual explanations. This technique involves showing how a model's output would change if its inputs were different. For example, a counterfactual explanation for a loan denial might be: "The loan would have been approved if the applicant's income were $10,000 higher." Counterfactuals can provide personalized, actionable insights, although they also raise ethical questions that we will explore later.

Lastly, we have interactive explanation tools. These software applications enable users to interact with AI models, changing their inputs and observing their outputs. These tools can facilitate a greater comprehension of AI, as users can learn through active exploration and experimentation.

These techniques are not panaceas; each has its strengths and limitations. Nor are they exhaustive—AI explainability is a vibrant research area with new tools and techniques constantly emerging. But together, they form a powerful toolbox for demystifying AI and transforming it from an inscrutable black box into a clear, comprehensible stream.

In this chapter, we will learn to use these tools effectively and overcome the ethical and practical challenges they pose. We will also learn from pioneers in AI explainability, drawing on their wisdom and experience to enrich our understanding. Together, we will explore the intriguing landscape of AI explainability, guided by the beacon of transparency.

BALANCING MODEL COMPLEXITY AND INTERPRETA-BILITY

In the depths of the AI forest, we face a choice between two paths: one that leads to complex models and another that leads to interpretability. It is a delicate dance, a balancing act of sorts, for, as we often find, the more complex a model becomes, the more difficult it is to interpret. How then do we navigate this precarious terrain, ensuring that we do not lose ourselves in the labyrinth of complexity while still harnessing the power of intricate AI models?

This balance between model complexity and interpretability has been a recurring theme in the evolution of AI. Early models were relatively simple and interpretable. Linear regression, for example, creates a clear and understandable relationship between input and output. Yet their simplicity often limited their predictive power.

As we ventured further into the forest, the models became more intricate. We encountered random forests, support vector machines, and deep neural networks, each more complex and powerful than the last. These models, particularly deep learning, have achieved astonishing results, often surpassing human performance. However, they have also become increasingly opaque, earning the moniker of black box models.

The crux of the issue lies in the trade-off between performance and interpretability. More complex models can capture nuanced patterns in data, improving performance. However, their inner workings become less comprehensible, making it hard to explain their decisions. On the other hand, simpler, more interpretable models can provide clear explanations but may fall short in performance.

Yet as we stand at this crossroads, we should not resign ourselves to a binary choice between complexity and interpretability. Instead, we should strive for synergy—a harmonious coexistence between the two. How can we do this? There are several approaches.

First, we can use model-agnostic methods like LIME and SHAP. These techniques allow us to glean insights from complex models without simplifying them. Second, we can design models that are inherently interpretable, such as decision trees or interpretable neural networks. Third, we can adopt an ensemble approach, combining various simple models to achieve comparable performance to a complex one. Lastly, we can use visualization techniques to make complex models more understandable.

As we navigate this path, we must also consider the context. In some cases, performance might be paramount, such as in cancer diagnosis, where even a slight improvement can save lives. In other cases, like credit scoring, interpretability might take precedence to ensure fairness and accountability.

The balance between model complexity and interpretability is not a tightrope but a bridge, one that we must traverse with careful consideration and contextual understanding. As we explore this balance throughout this chapter, the twin lights of performance and interpretability will serve as our guides. With curiosity and courage, we will proceed to unravel the mysteries of AI.

Expert insights and examples

In our exploration of transparency and explainability in AI, we find ourselves in a clearing—a gathering place where we can pause and learn from those who have traversed this terrain before us. We are not alone in this journey, and the insights from experts and the lessons from real-world case studies serve as our guiding stars.

We first learn from Dr. Cynthia Rudin, a renowned computer scientist and advocate for interpretable machine learning. Dr. Rudin shares her experiences from the Duke Computer Science prediction analysis lab, where her team has developed highly accurate yet transparent models for predicting electricity grid failures and diagnosing sleep apnea. She emphasizes the need for simplicity in machine learning models and points out that black box models are not always necessary or desirable, especially in critical applications where errors can have serious consequences.

Next, we learn from Dr. Rich Caruana, a senior principal researcher at Microsoft Research. Dr. Caruana's work on intelligible models for health care provides invaluable insights into the trade-off between model complexity and interpretability. His story of developing a pneumonia risk prediction model brings to life the dangers of opaque AI models. The neural network model initially suggested that asthmatic patients should be sent home, which seemed like an illogical choice. Upon inspection with a more interpretable model, the team discovered this was because asthmatic patients were usually directly sent to intensive care, making them appear "safer" in the data. This vital nuance would have been lost in a black-box model.

We also examine real-world case studies, like IBM's AI Explainability 360 Toolkit—an open-source library of algorithms that support the interpretability and explainability of data and machine learning models. We see how businesses like FICO use interpretable machine learning models to make decisions about credit risk that are both highly accurate and readily explainable to customers and regulators.

Eminent voices in the field of AI ethics sound alarms regarding the broad societal impacts of impenetrable artificial intelligence systems. They underscore the urgent need for interpretability and accountability across the AI landscape, especially in high-stakes sectors such as law enforcement and legal rulings. Without transparency in these contexts, AI risks amplifying engrained prejudices and inequities, potentially enabling oppression rather than justice. By spotlighting real-world implications, experts spur collective soul-searching into how we integrate AI into civic life. Their warnings make clear that intelligibility is not just a technical challenge but a moral imperative. Absent purposeful efforts to demystify AI, we cede control of our collective future to black-box algorithms tuned toward efficiency rather than ethics.

As we learn from these expert insights and dissect these case studies, we gain a greater understanding of the intricate dance between model complexity and interpretability. We begin to appreciate the nuances and complexities of ensuring transparency and explainability in AI, realizing that it is not just a technical challenge but also a societal one.

Armed with this newfound knowledge and understanding, we continue our journey, our path illuminated by the insights gleaned from these conversations and case studies. The road ahead is challenging, but we are better equipped to navigate the complexities and make more informed decisions in our quest for transparent and explainable AI.

Insights Examples

CHAPTER THREE

AI UNDER THE LAW: NAVIGATING AI LEGISLATION AND REGULATION

AI AND THE LAW: CURRENT AI-RELATED LAWS AND REGULATIONS

Our AI journey's third chapter takes us to the point where technology meets the law. The governance of artificial intelligence is a rapidly evolving area. It is a complex labyrinth that requires both technical expertise in AI and a broad understanding of legal, ethical, and societal implications.

The current landscape of AI-related laws and regulations is a patchwork of measures scattered across jurisdictions and varying in scope and substance. Due to the relentless pace of AI innovation and the growing awareness of its societal effects, it is a dynamic and fast-paced environment.

In the European Union, for instance, we see one of the most comprehensive attempts to regulate AI. The proposed Artificial Intelligence Act, introduced in 2021, seeks to establish a legal framework for "trustworthy AI." The act classifies AI systems based on their risk to fundamental rights and imposes corresponding regulatory requirements, ranging from transparency and human oversight for "high-risk" AI systems to outright bans on certain "unacceptable risk" applications, like AI-based social scoring.

In the US, regulations have been more targeted toward certain AI applications, such as autonomous vehicles, facial recognition, and health care. The Algorithmic Accountability Act 2022, for instance, proposes that companies conduct impact assessments of their high-risk automated decision systems.

In China, the approach to AI regulation reflects the country's broader sociopolitical context, with an emphasis on state control and a focus on harnessing AI for economic growth and social stability. The country's New Generation Artificial Intelligence Governance Principles, released in 2019, emphasize the importance of "controllable security" and the ethical use of AI.

These are merely snapshots of a global landscape with a variety of approaches and an ongoing discussion about the type and scope of AI regulation. In navigating this landscape, it is essential to understand not only the letter of the law but also its spirit—the underlying principles and values it seeks to uphold.

In this chapter, we'll look into how laws and regulations pertaining to AI are not just dry legal texts but rather dynamic, evolving entities. We will look beyond the mere provisions to the broader context, understanding how different societies grapple with the profound questions posed by AI—questions of privacy, fairness, accountability, and human dignity.

So let us embark on this journey, a journey not just of learning but of critical reflection and nuanced insight. The trail may be complicated and the road challenging, but with a curious mind and an open heart, we can solve this puzzle, gaining insights that will illuminate our understanding of AI and its place in our world.

Privacy, data rights, and liability issues in AI

Profound questions and heated debates are prevalent in the realm of artificial intelligence regulation. Among these, issues of privacy, data rights, and liability stand out as particularly contentious and complex. These issues touch on some of the most fundamental aspects of our lives and societies, raising questions about the very nature of personhood, dignity, and responsibility in the age of AI.

Privacy is perhaps the most visible and widely discussed aspect. As AI systems grow more powerful and pervasive, they can process vast amounts of information, often of a personal and sensitive nature. This ability, while fueling their efficiency and effectiveness, also raises serious concerns about privacy. From social media algorithms that seem to know our preferences better than we do to facial recognition systems that can identify us in a crowd, the reach of AI into our private lives can feel both uncanny and unsettling. This invasion of privacy is not just a subjective discomfort; it can lead to concrete harms such as identity theft, discrimination, and manipulation.

Data rights are intricately linked to privacy but extend beyond it. They refer to the rights individuals have over their personal details—rights to access, correct, delete, and transfer their data, as well as to object to or restrict its processing. In the context of AI, these rights take on new dimensions. For instance, how do we exercise our right to correct our data when it has been used to train an AI system? How do we transfer our data when it is embedded in a complex network of AI-enabled services? These questions pose significant challenges for both the design and regulation of AI systems.

Liability issues arise when AI systems cause harm. Traditional notions of liability are based on human agency and intent. But with AI, these notions become blurred. Who is accountable in the event that an autonomous vehicle causes an accident—the manufacturer, the software developer, the user, or the AI system itself? Similar questions arise in other domains, from health care to finance. Addressing these issues requires not only legal innovation but also a broader societal dialogue about responsibility and accountability in the age of AI.

As we move forward, we will explore the complexities and controversies of these issues. We will bring to light the tensions and trade-offs, the challenges, and the opportunities. We will study diverse voices—from legal scholars and technologists to privacy advocates and ordinary citizens—and weave their insights into a nuanced understanding of these crucial issues.

This road will lead us into uncharted territory, where the familiar landmarks of law and ethics give way to new and unsettling landscapes. But it is precisely in these moments of uncertainty and discomfort that we have the opportunity to rethink, reimagine, and reshape our relationship with AI. So let us press on, not with fear or resignation but with curiosity, courage, and a commitment to building a future where AI serves humanity, respects our rights, and upholds our dignity.

Envisioning the future of AI legislation

As we stand at the crossroads of human history, grappling with the power and promise of artificial intelligence, we are tasked with the monumental challenge of sketching the contours of future legislation that will govern this burgeoning field. It is a task that requires not just legal acumen but also technological savvy, ethical sensitivity, and a visionary perspective.

Technology advancements, societal values, economic considerations, and geopolitical dynamics, among other things, will all shape the future of AI legislation, which is likely to be as complex and multifaceted as AI itself. It is a future that demands our attention, our imagination, and our best efforts to balance the various interests at stake.

One key aspect of this future legislation will be the "regulation of data." As the lifeblood of AI, data will be at the heart of many legal debates and decisions. How do we protect individuals' privacy while enabling AI innovation? How do we ensure that data rights are respected and enforced, even as data flows across borders and sectors? And how do we deal with the challenges of anonymization, consent, and security in the age of big data?

Another crucial aspect will be the "governance of AI decision-making." As AI systems become more autonomous and influential, the question of who decides—and on what basis—will become increasingly important. How do we ensure transparency and accountability in AI decision-making? How do we manage the trade-off between explainability and performance in AI models? How do we deal with the risks of bias, discrimination, and manipulation in AI-driven processes?

A third area of focus will be "liability for AI-caused harm." As AI systems take on roles traditionally performed by humans, the traditional principles and practices of liability will need to be rethought. Should we stick to human-centric models of liability, or should we consider new models that account for the unique characteristics of AI? Should we hold AI systems themselves legally accountable, or should we focus on the humans behind these systems?

And beyond these specific areas, there will be broader issues to consider: How do we foster innovation and competition in the AI industry? How do we manage the global dimensions of AI, from cross-border data flows to international standards and cooperation? How do we ensure that AI benefits all segments of society, reducing rather than exacerbating inequalities?

We should investigate these inquiries and imagine potential responses, drawing on knowledge from law, technology, ethics, and social science. We should learn from leading experts and stakeholders, highlighting diverse perspectives and innovative proposals. And through this exploration, we will seek to chart a course toward a future where AI is not just powerful and efficient but also ethical, equitable, and accountable—a future where AI truly serves humanity.

CHAPTER 4:

SECTION 1:

AI IN THE GLOBAL SOUTH: CURRENT LANDSCAPE AND POTENTIAL

CHAPTER 4

CHAPTER CHAPTER 4

As we stand on the threshold of an age defined by artificial intelligence, it is crucial to consider the technological landscape beyond the traditional tech powerhouses. The Global South, comprising regions of Latin America, Africa, and developing Asia, offers unique opportunities and challenges for AI adoption and development. In this chapter, the current state of AI in the Global South is explored, exposing its exciting potential while recognizing the hurdles that must be overcome.

AI's current landscape in the Global South is an intriguing amalgam of burgeoning initiatives, innovative applications, and nascent development. In countries like Brazil, India, and South Africa, we see an expanding tech-savvy population harnessing AI in creative ways, from enhancing agricultural productivity to transforming healthcare delivery. Start-ups are mushrooming, leveraging AI to address local challenges and tapping into rapidly digitalizing economies.

Yet the landscape is not without its disparities. Access to AI technology is far from uniform, with a digital divide that threatens to exacerbate existing socio-economic inequities. High-quality data, essential for training robust AI models, is often lacking or inaccessible. Infrastructure deficits, educational gaps, and the dearth of policy frameworks further complicate the path to AI proliferation.

Despite these challenges, the potential of AI in the Global South is prodigious. AI has the potential to serve as a powerful tool for socio-economic development, tackling region-specific problems from disease control to disaster management. For instance, AI can optimize crop yields in agrarian economies, provide telemedicine services in remote areas, or enhance educational outcomes through personalized learning.

AI can also catalyze the Global South's journey toward the Fourth Industrial Revolution, fostering economic diversification, job creation, and global competitiveness. It can help countries leapfrog traditional stages of development, bypassing infrastructural limitations and resource constraints.

However, realizing this potential requires concerted efforts at multiple levels. Governments need to create enabling policy environments that foster AI innovation and adoption while mitigating risks. Businesses need to invest in AI capacities, embedding AI in their operations and offerings. Educational institutions have to nurture AI talent, equipping young minds with the necessary skills and knowledge.

We will traverse this multifaceted landscape, drawing on real-world examples, expert insights, and empirical studies. We will be diving into the exceptional use of AI in the Global South, analyzing the difficulties, and imagining methods to leverage AI's transformative potential. Through this journey, we aim to offer a nuanced and future-oriented perspective on AI in the Global South.

CHAPTER 4:

SECTION 2:

ETHICAL CONSIDERATIONS AND RISKS IN AI DEVELOPMENT IN THE GLOBAL SOUTH

CHAPTER 4

In the quest to harness AI's transformative potential, it is crucial not to lose sight of the ethical considerations and risks that such a powerful technology entails. As AI permeates various aspects of society, it brings to the fore complex ethical dilemmas and potent risks, particularly in the context of the Global South.

AI's ethical considerations stem from its inherent characteristics. Its decision-making process raises questions about accountability and fairness. Its reliance on data magnifies privacy concerns. Its potential for automation ignites fears of job displacement, and its capacity to influence behaviors stokes worries about manipulation and control.

In the Global South, these ethical considerations acquire unique dimensions. The digital divide can translate into an AI divide, with AI benefits accruing to the few who have access to technology. Data privacy concerns are magnified in regions with weak data protection laws. The risk of job displacement is particularly acute in economies reliant on sectors susceptible to automation, such as manufacturing and call centers.

AI also poses risks that can have far-reaching implications. One such risk is the reinforcement of existing biases and inequalities. AI systems, trained on biased data or designed without considering local contexts, can perpetuate stereotypes, discriminate unfairly, and exclude marginalized communities.

Another risk is the misuse of AI for surveillance and control. In the absence of strong checks and balances, governments and corporations can harness AI to monitor individuals, suppress dissent, and infringe on human rights. The Global South, with its varied political regimes and governance structures, is particularly vulnerable to such misuse.

Addressing these ethical considerations and mitigating these risks requires a multipronged approach. It calls for embedding ethics in AI design and development, adopting a human-centric approach that prioritizes fairness, transparency, and inclusivity. It necessitates robust policy frameworks that protect data privacy, prevent misuse, and ensure accountability. Furthermore, it also demands the active engagement of all stakeholders—from governments and businesses to civil society and citizens—in shaping AI's trajectory.

The forthcoming sections will examine these ethical considerations and risks. We will examine their manifestations in the Global South, drawing on real-world examples and case studies. We will also explore strategies to address these issues, offering a roadmap for ethical, responsible, and inclusive AI development in the Global South.

CHAPTER 4:

SECTION 3:

CASE STUDIES: AI ADDRESSING REGION-SPECIFIC PROBLEMS IN THE GLOBAL SOUTH

The practicality of AI as a tool for change truly shines when we examine its applications within specific regional contexts. It becomes apparent that AI has the potential not only to transform but also to uplift societies, particularly in the Global South. Let us examine insightful case studies that exhibit how AI has been utilized to overcome region-specific hurdles.

Case study 1: AI for agriculture in India

In India, where over half the population depends on agriculture for their livelihood, small-scale farmers grapple with challenges like uncertain weather patterns, pest attacks, and fluctuating market prices. Enter PEAT, a German start-up that developed Plantix, an AI-based mobile application. It helps farmers identify plant diseases and pests just by analyzing photos taken with their smartphones. Additionally, the app provides advice on how to treat these issues and alerts about potential risks, enabling farmers to increase their crop yield and reduce their dependence on costly pesticides.

Case study 2: AI for health care in Rwanda

Rwanda, a small country in East Africa, has leveraged AI to revolutionize its healthcare system. In partnership with the American company Zipline, the

Rwandan government has used AI-powered drones to distribute medical supplies, including vaccines and blood, to remote areas of the country. This has significantly reduced the time taken to provide critical medical supplies, saving countless lives.

Case study 3: AI for education in Brazil

In Brazil, the educational tech start-up Geekie uses AI to personalize learning for students. Its platform, Geekie Games, uses machine learning algorithms to adapt to each student's learning pace and style, offering tailored exercises and study plans. Millions of students across the country have used this tool, which helped them prepare for university entrance exams and reduced the educational gap.

These case studies offer a glimpse into the transformative potential of AI in the Global South. They underscore the importance of developing and implementing AI solutions that are sensitive to local contexts, inclusive, and designed to benefit the many, not just the few. Moving ahead, the task is not just to utilize the potential of AI but to guarantee that its advantages are evenly spread, risks are reduced, and ethical considerations guide its implementation.

CHAPTER FIVE

05

SECTION ONE:

The Environmental Cost of AI: Understanding the Carbon Footprint of AI Models

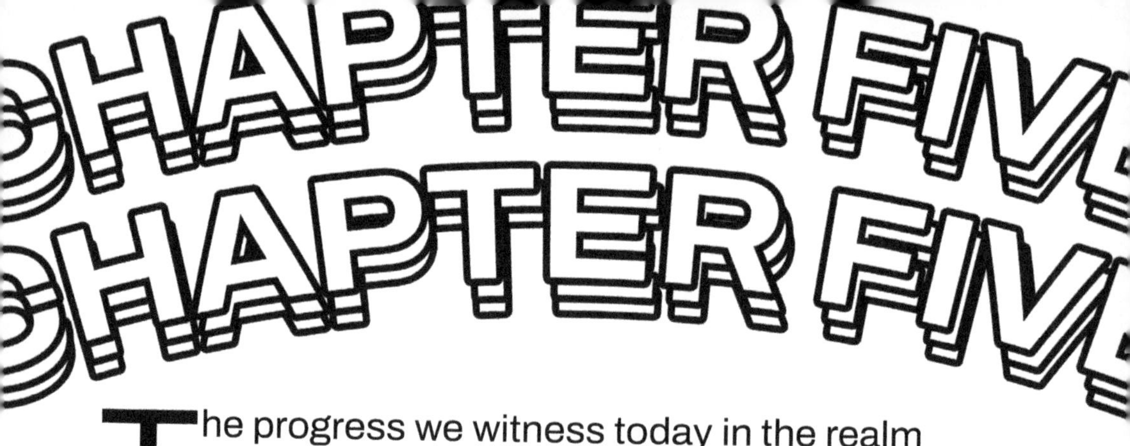

The progress we witness today in the realm of AI—from self-driving cars to speech recognition—is exciting and truly revolutionary. Nevertheless, these advancements have an impact on the environment that is often overlooked. Yes, the computational powerhouses driving AI advancements are energy hogs, and this has implications for our planet.

To comprehend the connection between AI and the environment, understanding the notion of a carbon footprint is essential. In the simplest terms, a carbon footprint is the total amount of greenhouse gases, in particular carbon dioxide, which are vented into the atmosphere as a result of human activities. In the context of AI, this refers to the emissions that result from data centers and computing infrastructure powering AI models.

Training complex AI models involves heavy computational workloads. For example, a machine learning model must process vast amounts of data, typically requiring the equivalent computational power of thousands of personal computers running for days or even weeks. This energy consumption leads to substantial carbon dioxide emissions.

Consider the case of OpenAI's development of GPT-3, an advanced language model. Training this single model reportedly consumed the same amount of electricity as the average American household would use over a century. The carbon emissions associated with this energy use are comparable to those of a transatlantic flight.

The demand for energy is getting worse because of the trend of creating bigger models with only slight performance enhancements. The pursuit of state-of-the-art results in the AI field often overlooks the environmental cost of scaling up these models.

In an era where climate change is one of the most formidable dilemmas facing humanity, it is essential to bring the environmental impact of AI into the limelight. We must ensure that the growth of this transformative technology does not come at the expense of our planet's health. The next step is to explore how we can reduce this carbon footprint, make AI models more energy-efficient, and balance the race for AI innovation with our shared responsibility for environmental sustainability.

ARBON FOOTPRINTS
ARBON FOOTPRINTS

SECTION TWO:

Toward a Greener AI: Strategies and Research Trends in Energy-Efficient AI

CHAPTER FIVE

Given the environmental impact of AI, we must ask ourselves: How can we create a future where AI and environmental sustainability are aligned? The solution lies in our collective push toward "greener AI," a term that signifies a shift to more energy-efficient AI models and practices.

The first step toward this goal is increasing the energy efficiency of the hardware used in AI computations. This involves a collective push from chip manufacturers, data centers, and AI researchers to innovate and implement energy-saving technologies. For instance, the advent of specialized hardware like tensor processing units (TPUs) and graphics processing units (GPUs) has already shown a significant reduction in energy consumption compared to traditional central processing units (CPUs).

However, hardware is just one piece of the puzzle. There is a growing body of research focusing on developing energy-efficient AI algorithms. The idea is to create AI models that can learn more efficiently, either by requiring less data, fewer computations, or both. Techniques such as model pruning, quantization, and knowledge distillation are being extensively researched in this context.

For instance, model pruning is a process where the less important connections in a neural network are eliminated, making the model leaner and faster without a significant loss of accuracy.

In addition to these, there is a push toward more accurate carbon accounting in AI research. This would involve researchers and institutions quantifying and reporting the carbon emissions associated with their AI models, akin to how scientists quantify and report the energy use or performance of their models today. Such transparency can help create industry-wide awareness and drive competitive innovation towards greener AI.

Lastly, we must also look beyond the models and algorithms and reconsider the culture of AI development. The current norm of continuous model upgrading and deployment has significant environmental implications. Embracing a more thoughtful and intentional approach to AI development—one that considers the trade-offs between performance gains and environmental costs—is essential.

We will explore these strategies in the upcoming sections, along with their practical uses and their impact on the future of AI. As we navigate these narratives, it becomes clear that the path toward a greener AI is not just a necessity but a possibility within our reach.

SECTION THREE:

Case Studies: Pioneering the Path to Greener AI

CASE STUDIES

CASE STUDIES

While we deal with the effects of AI's carbon footprint, many groups and organizations are working to build an eco-friendlier AI environment. Let us dive into a few interesting case studies that show the innovative ways in which this problem is being tackled.

Case study 1: Google's energy-efficient TPUs

1

Google has been at the forefront of building energy-efficient AI hardware. Their tensor processing units (TPUs), custom-designed to accelerate machine learning workloads, have helped reduce Google's energy usage. The third generation of TPUs, for instance, delivers up to 420 teraflops of computing power and uses a liquid cooling system, which is more energy-efficient than traditional air cooling. The fourth generation outperforms the third by 2.1 x on average on a per-chip basis and improves performance.

Case 2: OpenAI's model pruning techniques

OpenAI, a leading AI research organization, is actively exploring techniques such as model pruning to reduce the computational cost of AI. Their work has shown that it is possible to reduce the size of language models like GPT-3 by up to 90% without a substantial decline in performance. By adopting this leaner model, not only will it reduce energy consumption, but it will also enhance accessibility for users with limited computational resources.

Case 3: DeepMind's AI for Google data centers

Alphabet's AI company, DeepMind, has utilized machine learning to optimize the cooling of Google's data centers, where thousands of servers run services like Search, Gmail, and YouTube. Their AI system reduced the energy used for cooling by up to 40 percent, demonstrating how AI itself can be a tool for energy efficiency.

Case study 4: the MLPerf Consortium's push for carbon accounting

The MLPerf Consortium, a group of companies and research institutions, has initiated a move to include energy consumption as part of their AI benchmarking suite. By doing so, they hope to drive a culture of carbon accounting in AI research, encouraging researchers to innovate not just for performance but also for energy efficiency.

Through these case studies, we see the exciting potential of greener AI unfold. They highlight the multidimensional nature of the challenge at hand—requiring advancements in hardware, software, and cultural norms alike. As we continue our journey toward more sustainable AI, these case studies serve as beacons, lighting the path toward a future where AI and environmental sustainability coexist.

CHAPTER

CHAPTER SIX CHAPTER
HAPTER SIX CHAPTER
CHAPTER SIX CHAPTER

section 1:

Understanding Human-AI Collaboration: Defining What Human-AI Collaboration Means and Why It is a Vital Area of Focus in the AI Field

APTER SIX CHAPTER S
CHAPTER SIX CHAPTER

The more we explore the intricate domain of artificial intelligence, the more apparent it becomes that the best route to triumph lies in combining man and machine. This brings us to the concept of human-AI collaboration, a rising paradigm in the field that seeks to leverage the unique strengths of both entities for enhanced outcomes.

Human-AI collaboration, often termed "augmented intelligence," is a model wherein humans and AI systems work together, each contributing their distinct expertise to achieve common goals. This approach is not about machines replacing humans but about AI enhancing human decision-making and capabilities.

Why is human-AI collaboration such a vital area of focus in the AI field? The answer lies in the inherent strengths and weaknesses of both human beings and AI systems. Humans, on the one hand, bring to the table a nuanced understanding of the world, the ability to perceive context, a capacity for creative thinking, and the power to make ethical and moral judgments. Machines, on the other hand, excel at processing vast amounts of data, performing repetitive tasks with unfailing accuracy, and making complex calculations at lightning speed.

By integrating these complementary skills, we can create a powerful symbiosis that enhances our capabilities. For instance, in health care, AI algorithms can analyze millions of medical images to detect potential anomalies, while human doctors provide the final diagnosis, considering the patient's overall health, history, and context that the AI may not fully grasp.

In business, while AI can churn through enormous datasets to generate actionable insights, human judgment is crucial in strategic decision-making, where context, ethics, and long-term vision come into play. However, realizing the full potential of human-AI collaboration is not without challenges. We must address issues of trust, transparency, and control by defining clear boundaries for AI systems while ensuring they are user-friendly and understandable for those who use them. As we continue to advance AI, we must constantly recalibrate this balance, ensuring we are creating systems that not only augment human intelligence but also respect human autonomy.

The subsequent sections of this chapter will cover the techniques for fostering successful human-AI teamwork, the ethical issues involved, and the game-changing potential it has in different industries. As we navigate this promising frontier, it is clear that the future of AI is not about humans versus machines but humans with machines.

CHAPTER SIX

section 2:

Augmenting Human Capabilities: Exploration of How AI Can Augment Human Capabilities Rather Than Replace Them

The most transformative promise of artificial intelligence lies not in replacing human intelligence but in amplifying it. This augmentation allows us to transcend our cognitive limitations, thereby helping us to make more informed decisions, solve complex problems, and broaden our creative horizons. Let us begin a detailed exploration of how AI augments human capabilities across various sectors.

Health care

In health care, AI is revolutionizing diagnosis, treatment, and patient care. Machine learning algorithms can analyze vast medical datasets, identifying patterns and anomalies far beyond human capability. Radiologists, for instance, can use AI to detect early-stage cancers in medical imaging, with systems trained to recognize the most minute abnormalities. However, AI does not replace the physician. Instead, it serves as a robust tool, augmenting their diagnostic capability and allowing them to make more accurate, timely diagnoses.

Business

In the business realm, AI provides strategic advantages by enhancing human decision-making. AI algorithms can assess extensive volumes of data to uncover trends, predict customer behavior, and optimize operations. Yet the human role remains pivotal for strategic and ethical decisions, interpreting AI-generated insights, and understanding their broader implications within specific contexts. Thus, AI acts as a force multiplier, enabling business leaders to make informed decisions swiftly and confidently.

Education

Education is another domain where AI significantly augments human capabilities. AI-powered personalized learning systems adapt to each student's unique learning style and pace, providing customized educational content. Teachers use these insights to better understand each student's progress and tailor their teaching methods accordingly. Here, AI does not substitute the teacher; instead, it equips them with better tools to fulfill their role.

Creative Industries

Even in creative industries, where human ingenuity has always reigned supreme, AI is playing an increasingly supportive role. Artists and designers are integrating AI tools to generate new ideas, explore different creative paths, and even critique their work. However, the final creative decisions and the emotional resonance that define great art still come from the human creator. In this context, AI serves as a collaborative partner, expanding the artist's creative landscape.

In each of these sectors, AI's role is not to supplant humans but to amplify our capabilities, enabling us to achieve more than we could alone. As we embrace this augmented intelligence, we must also tread carefully, ensuring that AI systems are transparent, controllable, and designed to respect human autonomy and dignity. As we will discuss in the following sections, managing this delicate balance is one of the paramount challenges of our AI-augmented future.

section 3:

The Future of Work with AI

The future of work in the age of artificial intelligence invokes a complex narrative filled with excitement, curiosity, and a hint of apprehension. As we navigate through this developing narrative, one thing is clear: the job landscape is shifting, and AI is at the helm of this change.

Artificial intelligence holds the potential to automate a substantial proportion of mundane, repetitive tasks, irrespective of the industry. This shift is not just about manual labor; even cognitive tasks such as data entry, accounting, and scheduling can be handed over to AI systems. As we offload these tasks to AI, we free up human capacity for tasks that truly require our unique human attributes.

Consider the field of customer service, for example. Artificial intelligence can automate responses to simple queries, allowing customer service representatives to prioritize on complex issues that demand empathy, understanding, and nuanced problem-solving—qualities that AI is yet to master.

In health care, AI can take over repetitive tasks like maintaining patient records, coordinating appointments, or even interpreting specific diagnostic tests. This automation allows healthcare professionals to devote more time to direct patient care and complex decision-making.

In the corporate world, AI can oversee a multitude of administrative tasks, enabling employees to focus on strategic thinking, collaboration, and innovation. By taking over routine tasks, it empowers humans to dig deeper into their creative and strategic capacities.

This shift toward complex and creative tasks could potentially revolutionize our perception of work. We will no longer be "human doings," performing tasks in a mechanical fashion. Instead, we will grow into "human beings," where our work becomes an expression of our creativity, empathy, strategic thinking, and our ability to connect with others.

However, as we envision this promising future, we must also address the challenges that accompany it. These challenges include the risk of job displacement due to automation and the urgent need for reskilling and upskilling the workforce. It is crucial that we invest in education and lifelong learning, enabling individuals to adapt to the changing job landscape. We need to ensure that the benefits of AI-driven automation are equitably distributed. Policymakers, business leaders, and educators must come together to create a roadmap that ensures everyone benefits from AI's transformative potential, minimizing displacement and inequality.

As we stand on the brink of this transformation, we have the unique opportunity to redefine the nature of work, fostering a future where work is not just about survival but also about growth, fulfillment, and the expression of our uniquely human capabilities. Let us seize this opportunity to shape an AI-augmented future of work that is fair, fulfilling, and truly human-centered.

section 4:

Case Studies: Real-World Instances of Successful Human-AI Collaborations

In the real world, artificial intelligence has effectively collaborated with humans to enhance efficiency, creativity, and innovation. Let us explore some of these fascinating case studies.

Health care: PathAI 1

PathAI, a Boston-based start-up, has made significant strides by collaborating with pathologists. Their AI-based tool aids in diagnosing diseases from pathology slides, providing an accuracy rate that surpasses that of human-only teams. However, the system does not replace pathologists. Instead, it augments their capabilities by reducing error rates and increasing efficiency, allowing the pathologists to focus on complex diagnoses.

Education: CENTURY Tech

In the education sector, CENTURY Tech's AI platform has transformed personalized learning. By analyzing a student's learning behavior, the AI tailors educational content, helping students learn at their own pace and style. This tool does not replace teachers but empowers them. Armed with insights about each student's learning style, teachers can better guide their students, focusing on areas where human intervention is most valuable.

3 Agriculture: Blue River Technology

Blue River Technology's targeted weed control is a remarkable example of human-AI collaboration. This system uses machine learning algorithms to differentiate between crops and weeds, spraying herbicides only on the latter. This AI-driven approach has reduced herbicide usage by up to 90 percent, contributing to more sustainable farming. Here, AI takes over a laborious task, enabling farmers to focus on broader farm management strategies.

Media: the Washington Post's Heliograf

In journalism, the Washington Post's AI technology, Heliograf, has automated the creation of short news updates, particularly for sports and election coverage. This automation has not made journalists redundant. On the contrary, it has freed them from repetitive reporting, allowing them to focus on investigative journalism, feature articles, and in-depth analyses where their human expertise is most needed.

Retail: Stitch Fix 5

Stitch Fix, an online styling service, uses AI to tailor clothing selections for customers based on their past preferences, sizes, and feedback. However, the AI does not make the final decision. Human stylists review the AI's recommendations, adding their personal touch before finalizing each shipment. In this way, Stitch Fix combines AI's scalability and precision with human intuition and creativity.

These case studies offer a glimpse into the future of human-AI collaboration—a future where AI does not replace humans but augments our capabilities. By automating repetitive tasks, AI empowers us to channel our creative, strategic, and empathetic capacities into areas where we truly make a difference. The rise of AI is not a threat but an opportunity to create a more fulfilling, creative, and human-centered future of work.

section 5:

Challenges and Opportunities in Human-AI Collaboration

Human-AI collaboration holds tremendous potential, but it is not without its share of challenges. Trust, communication, and decision-making stand out as significant hurdles. However, it is important to remember that every challenge presents an opportunity for growth and innovation. Let us explore these areas further.

1 Trust in AI

One of the most prevalent challenges is fostering trust between humans and AI. AI's decisions often occur within a black box, making it difficult for humans to comprehend and trust the process. Moreover, high-profile failures of AI systems, such as autonomous vehicle accidents, can further undermine trust.

However, this challenge presents an opportunity to improve transparency and explainability in AI systems. By developing models that can provide understandable reasons for their predictions and actions, we can foster trust and enable effective human-AI partnerships. Research into explainable AI (XAI) is a burgeoning field addressing this need.

2 Communication between humans and AI

Effective communication is another challenge in human-AI collaboration. Humans and AI systems typically speak different languages, with humans using ambiguous and context-dependent language, while AI systems operate best with precise and unambiguous instructions.

This challenge underscores the opportunity to improve natural language processing (NLP) capabilities in AI systems. The communication gap is closing thanks to advancements in NLP, as seen in models like GPT-4, which make it easier for AI systems to comprehend and respond to human language. Further refinements could lead to even more natural and efficient human-AI interactions.

3 Decision-making in AI

Decision-making in AI presents another challenge. While AI systems can analyze vast amounts of data and make decisions rapidly, they lack the human capacity for intuition, common sense, and ethical judgment. This can lead to decisions that, while technically correct, may be inappropriate or unethical in a broader context.

This challenge highlights the opportunity to integrate human oversight into AI decision-making processes. By creating systems where humans and AI collaboratively decide, we can combine the strengths of both. AI and humans can collaborate to make effective and ethical decisions. The former can handle the data-heavy analytical component, while the latter can provide intuitive, ethical, and contextual insights.

While challenges exist in human-AI collaboration, they also present opportunities for innovation and improvement. By addressing these challenges head-on, we can pave the way for a future where humans and AI work together seamlessly, each amplifying the other's strengths and mitigating their weaknesses. This is a future where AI does not replace humans but enables us to achieve more than we could alone.

section 6:

Conclusion and Future Directions

As we draw this exploration of human-AI collaboration to a close, it is essential to look forward and anticipate the trends and directions that will shape this exciting field. Our relationship with technology is at a crucial turning point, where AI is not only serving us but also working alongside us to enhance our abilities in unprecedented ways.

1 More intuitive human-AI interaction

A promising trend is the development of more intuitive and natural interfaces for human-AI interaction. Voice and gesture interfaces, sentiment analysis, and AI systems capable of understanding and adapting to individual users' unique styles of communication are all areas of active research and development. The AI of the future will not only understand our commands but also our context, emotions, and unspoken needs, creating a more seamless and intuitive experience.

2 AI as a creative partner

AI's role in creative endeavors is poised to expand. We have already seen AI generate music, art, and even literature. However, the future lies not in AI replacing human creativity but in augmenting it. Imagine AI tools that can understand and even anticipate a musician's creative vision or an AI writing assistant that can seamlessly continue an author's narrative in their unique style. Such AI systems would not replace human creatives but become creative partners, enabling new forms of expression and innovation.

3 Collaborative decision-making

In sectors from health care to finance, we will see more AI systems designed to collaborate with humans in decision-making. These systems will offer the best of both worlds: the data-crunching power of AI coupled with the ethical, intuitive judgment of humans. In areas like medicine, AI could analyze complex patient data, but human doctors would make the final decisions, taking into account the patient's unique circumstances and values. This has the potential to be revolutionary.

4 Democratization of AI

Finally, we can anticipate a trend toward the democratization of AI. As AI becomes more user-friendly and accessible, more people, even those without technical training, will be able to use and benefit from it in their work and daily lives. This could lead to a surge of grassroots innovation as people from diverse fields find novel uses for AI.

The future of human-AI collaboration is rich with possibilities. It is a journey of mutual growth and learning where humans and AI systems will continually adapt to and learn from each other. As we move forward into this exciting future, let us do so with a sense of curiosity, openness to innovation, and a firm commitment to ethical and fair practices. The AI of the future will not be something that happens to us, but something we shape together in partnership with the very AI systems we are creating.

CHAPTER

SECTION 1:

THE CURRENT STATE OF AI IN EDUCATION: SUCCESSES AND CHALLENGES

Artificial intelligence's foray into education has stirred a sea of transformative possibilities while also unveiling significant challenges. We will explore how AI is used in education, covering its achievements, difficulties, and the relationship between human educators and AI tools in shaping the future of learning.

SUCCESSES

AI has been instrumental in making education more personalized, accessible, and flexible. The traditional "one-size-fits-all" approach is giving way to customized learning experiences, where AI-powered platforms adapt to individual learners' pace, style, and level of understanding.

For instance, intelligent tutoring systems (ITS) like Carnegie Learning MATHia and Third Space Learning are revolutionizing personalized education. These platforms can gauge a student's understanding, provide customized feedback, and even change their teaching strategy in real time based on the learner's progress.

AI is also enhancing accessibility. Tools like Microsoft's Immersive Reader utilize AI to read text aloud, break words into syllables, and increase spacing between lines and letters, making learning more inclusive for students with dyslexia or visual impairment.

Additionally, AI has improved education's adaptability and flexibility, which is significant in the COVID-19 pandemic-driven era of remote learning. Artificial intelligence-powered platforms can provide 24/7 learning support, making quality education accessible beyond the confines of classrooms and traditional school hours.

CHALLENGES

Despite these successes, the application of AI in education faces substantial challenges. One of the primary concerns is data privacy. The very personalization that AI offers requires collecting and analyzing vast amounts of data on students, potentially infringing on privacy rights and exposing sensitive information to cybersecurity threats.

The issue of AI fairness also looms large. Biases in AI algorithms can inadvertently reinforce societal inequalities, such as when an AI system unintentionally favors learners from certain demographic backgrounds because of the data it was trained on.

There is a risk of overreliance on AI, leading to a diminished role for human teachers. While AI can provide personalized feedback, it currently cannot match a human teacher's ability to inspire, motivate, and understand the nuanced emotional and social aspects of learning.

Finally, there is the digital divide—the gap between those who have access to technology and those who do not. The benefits of AI in education are moot for students who lack access to reliable internet or the necessary hardware, a problem that is particularly acute in underserved communities and developing regions.

While AI has ushered in significant advancements in education, we must navigate these challenges with care. Balancing the benefits of AI with the need for data privacy, addressing algorithmic bias, maintaining the irreplaceable human touch in education, and ensuring equitable access to AI resources are key to harnessing AI's full potential in transforming learning.

SECTION 2:

THE FUTURE OF EDUCATION: AI FOR PERSONALIZED LEARNING AND STUDENT ENGAGEMENT

Moving ahead, it is crucial to realize that AI advancements will increasingly blend with the future of education. The road ahead for AI in education is one that promises greater personalization, increased student engagement, and a truly learner-centric approach.

AI AND PERSONALIZED LEARNING

AI's ability to customize educational content based on individual learning styles, pace, and preferences will further transform education into a personalized experience. AI systems will become more adept at diagnosing learner strengths and weaknesses, providing tailored educational content, and recommending personalized learning pathways.

Consider an AI-driven system that builds a dynamic learning profile for each student. It adjusts the pace of instruction, provides resources, and even suggests breaks when it detects a student's waning attention. As the system learns more about the student, it continually refines its approach, creating a truly personalized and adaptive learning journey.

AI FOR STUDENT ENGAGEMENT

Enhancing student engagement is another promising area for AI. By incorporating gamified elements, virtual reality, and interactive scenarios, AI can make learning more immersive, fun, and engaging. Imagine history lessons where students can virtually walk through ancient civilizations or science lessons where they can manipulate virtual molecules.

AI can also foster better student-teacher interactions. Teachers, freed from routine tasks by AI automation, can spend more time interacting with students, addressing complex concepts, and facilitating discussions that require higher-order thinking.

THE HUMAN-AI PARTNERSHIP IN EDUCATION

In envisioning this AI-augmented future of education, it is important to remember that AI is a tool that serves to enhance, not replace, the human element in education. The future of education lies in a harmonious partnership between AI and educators. AI can handle data analysis and personalization, while human teachers provide the empathy, social understanding, and inspirational mentoring that AI cannot replicate.

By using AI, struggling students can be identified, and teachers can offer timely and targeted support. It can free teachers from administrative tasks, enabling them to focus on student interaction and holistic development. And it can provide teachers with insights derived from data, helping them refine their teaching strategies.

CONCLU

While the journey to this AI-
challenges, from data privacy t
benefits to students and educa
navigate this path, it is essentia
education, one that can augme
to focus on what truly matters in
creative, and compassionate lea

SION

nabled future has its share of
algorithmic bias, the potential
rs are immense. As we
o approach AI as a partner in
our capabilities and free us
ducation: nurturing curious,
ners.

SECTION 3:

ETHICAL CONSIDERATIONS AND POTENTIAL RISKS

As we move toward an AI-enabled future in education, it is crucial to recognize and carefully manage the ethical considerations and potential risks involved.

DATA PRIVACY AND SECURITY

The foundation of AI's potential in education rests on its ability to process vast amounts of data. However, this raises significant concerns about data privacy and security. How can we ensure the protection of sensitive student data? This question becomes even more pressing as we consider the global differences in data protection regulations and the potential for misuse of data.

These concerns extend beyond the collection and storage of data. They encompass how data is used and who has access to it. For instance, the use of predictive analytics in education could potentially lead to profiling, where assumptions made by algorithms could unfairly influence a student's educational trajectory.

ALGORITHMIC BIAS

Another ethical concern arises from algorithmic bias. If the data used to train AI systems reflects societal biases, the AI could unintentionally perpetuate or even exacerbate these biases. For example, an AI system might favor students from certain backgrounds because of biased data inputs, potentially influencing the allocation of resources and opportunities.

EQUITY AND ACCESSIBILITY

While AI holds the potential to democratize education, it could also unintentionally widen the digital divide. Not all students have access to the same level of technology, and reliance on AI-enhanced education could leave those in technologically underserved areas at a disadvantage.

THE ROLE OF TEACHERS

The increasing use of AI in classrooms also questions the role of teachers. While AI can free teachers from administrative tasks and personalize learning for students, it is important to ensure that it does not inadvertently devalue or replace the irreplaceable human connection and mentorship that teachers provide.

NAVIGATING THE FUTURE

To navigate these ethical considerations and potential risks, it is critical to adopt a thoughtful and proactive approach. This involves establishing robust data protection policies, investing in the development of fair and transparent AI systems, ensuring fair access to AI-enhanced education, and continually reassessing the role of AI in the classroom.

By doing so, we can ensure that AI serves as a tool that enhances and supports education rather than an unchecked force that introduces new risks and inequalities. As we chart the course toward the future, let us ensure that we guide our journey with the principles of fairness, inclusivity, and respect for the human dignity of every learner.

SECTION 4:

CASE STUDIES

To bring our discussion on AI in education to life, let us explore two real-world case studies. These stories underscore the potential of AI to revolutionize pedagogy while also highlighting the challenges and ethical considerations we must navigate.

CASE STUDY 1: AI IN PERSONALIZED LEARNING—THE KNEWTON EXPERIENCE

Knewton, an adaptive learning technology provider, offers a prime example of AI's potential to enhance personalized learning. Knewton's system dynamically adapts to each student's unique needs, delivering customized content that optimizes their learning experience.

The software analyzes students' performance, identifies their strengths and weaknesses, and adjusts the content accordingly in real time. This personalized approach has led to notable improvements in student engagement and outcomes, with one study revealing a 19 percent increase in pass rates and a 12 percent rise in grades.

However, Knewton's use of AI also raises important ethical considerations. Concerns have been raised about the collection and use of student data, prompting Knewton to develop stringent data privacy policies to protect student information.

CASE STUDY 2: AI FOR STUDENT ENGAGEMENT—THE GEORGIA STATE UNIVERSITY EXPERIMENT

Georgia State University leveraged AI to boost student engagement and success rates. They developed an AI-powered chatbot named "Pounce" designed to answer students' questions and guide them through the enrollment process.

The results were staggering. The chatbot answered over 200,000 questions in its first summer, reducing "summer melt" (students who accept admission but do not enroll) by 22 percent.

While Pounce was a success, it also underscored the importance of human oversight. Initially, the chatbot provided incorrect answers to some questions due to misinterpretation of the queries. Georgia State University had to implement a feedback loop with human involvement to correct these errors, underscoring the fact that AI in education should augment, not replace, human involvement.

These case studies provide convincing examples of the transformative potential of AI in education. However, they also underscore the importance of approaching AI with a nuanced understanding of its ethical implications and potential risks. By doing so, we can harness the power of AI to enhance education while ensuring that we protect the interests of students and uphold our ethical obligations.

8

SECTION 1

The Imperative of Responsible AI Development

In the rapid march of technological progress, artificial intelligence has emerged as a formidable force, transforming every facet of human life. The promise of AI is, however, tempered by significant risk, as with all strong tools. The call for responsible AI development is not merely a matter of ethical niceties but a profound and urgent necessity.

Responsible AI development begins with an acknowledgement of AI's dual nature. AI systems, whether they are advanced machine learning algorithms or simple chatbots, can be powerful enablers of human progress. They can drive productivity, unlock new knowledge, and solve complex problems. However, these systems can also be used to manipulate, deceive, and infringe upon our privacy. The same algorithm that can predict disease can also be used to create biological agents, create deep fakes, manipulate reality, and spread disinformation.

Given this dual nature, responsible AI development hinges on the principles of integrity, fairness, and transparency. These principles should not be afterthoughts; they should be integral aspects of the design, development, and deployment of AI systems.

Transparency implies that AI systems should be understandable to human users. We must strive to eliminate the black box phenomenon and ensure that the workings of AI systems can be inspected and interpreted.

Accountability underscores that the consequences of AI systems should be attributable to identifiable entities. This accountability is not limited to the period of design and deployment but extends throughout the lifecycle of the AI system.

Fairness demands that AI systems not perpetuate existing biases or create new ones. They should treat all users fairly, and their benefits should be accessible to all.

Let us not forget that AI systems are human creations, reflecting our values, biases, and aspirations. The development of AI should be a participatory process involving a diverse group of stakeholders. This includes not just AI specialists but also ethicists, sociologists, and representatives of the communities in which these systems will have an impact.

At this point, it is worth emphasizing that the imperative of responsible AI development extends beyond the realm of AI practitioners. Policymakers, educators, and the public also have crucial roles to play. Policymakers must craft regulations that promote responsible AI practices; educators must instill a sense of responsibility in the next generation of AI practitioners; and the public must stay informed and hold AI systems and their creators accountable.

STAKEHOLDER ENGAGEMENT

The spectrum of stakeholders in AI is vast, encompassing users, communities, developers, businesses, regulators, and society at large. Identification and mapping are the first steps in effective stakeholder engagement, which then involves consistent communication, collaboration, and involvement.

The guiding principle here is inclusivity, which means giving voice to diverse perspectives, experiences, and fields of study. Engaging with nontechnical stakeholders and those whom AI systems might indirectly affect is part of this.

Feedback loops are also crucial. They facilitate the continuous flow of information, allowing developers to understand stakeholder concerns, expectations, and suggestions and incorporate this feedback into the AI development process.

ETHICAL CONSIDERATIONS

ETHICAL CONSIDERATIONS
ETHICAL CONSIDERATIONS
ETHICAL CONSIDERATIONS

Embedding ethics into AI development should be non-negotiable. This includes adhering to upright principles such as transparency, fairness, privacy, and accountability and operationalizing these principles in every stage of the AI life cycle.

Ethical considerations also extend to the question of "should," not just "can." Just because an AI system can be built does not mean it should be. Developers should consider the potential misuse and negative effects of the AI system and whether these risks outweigh the benefits.

LONG-TERM IMPACT ASSESSMENTS

LONG-TERM IMPACT ASSESSMENTS

Assessing the long-term effects of AI systems is a complex but necessary endeavor. This involves not just predicting the direct impacts of the AI system but also the indirect and systemic consequences.

To accomplish this, we need to adopt a systemic perspective, considering the interactions between the AI system and the broader sociotechnical system in which it operates.

We also need to think in terms of scenarios, not predictions. The future is inherently uncertain, and rather than trying to predict a single outcome, it is more useful to consider a range of possible scenarios and prepare for them.

Long-term impact assessments should be revisited and updated regularly as the AI system develops and as we learn more about its impacts.

In conclusion, stakeholder engagement, ethical considerations, and long-term impact assessments are pillars of responsible AI development. By adopting these best practices, we can ensure that our AI systems are not just technically sound but also socially beneficial and ethically aligned. This is a challenging endeavor, but one that we must embrace wholeheartedly as we navigate the AI landscape and shape the future of this powerful technology.

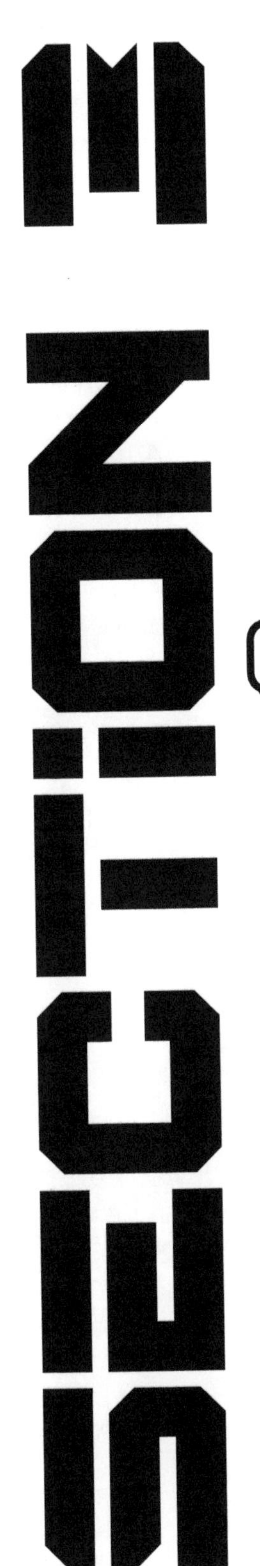

Case Studies of Responsible AI Development and Deployment

Case studies provide concrete examples of the principles we have discussed, demonstrating how responsible AI development and deployment can be achieved in practice. They also shed light on the challenges encountered along the way and how they can be overcome.

Case study 1: Google's AI principles in action

In June 2018, Google announced its AI principles, a set of commitments to guide the company's approach to AI. These principles include socially benefiting, avoiding creating or reinforcing unfair bias, building and testing for safety, being accountable to people, incorporating privacy design principles, upholding high standards of scientific excellence, and making available uses that accord with these principles. These principles have been updated several times, with the latest update in 2022.

A tangible implementation of these principles was seen when Google decided not to renew a contract with the US Department of Defense for Project Maven, which involved using AI to analyze drone footage. Google employees and external stakeholders voicing concerns influenced this decision, demonstrating a commitment to the principle of avoiding harmful applications of AI.

Case study 2: IBM's trustworthy AI

IBM has been a pioneer in the development of AI and has also led the way in defining principles for trust and transparency in AI. IBM's AI ethics guidelines emphasize the importance of transparency, explainability, fairness, and robustness.

IBM has operationalized these principles through initiatives like AI Fairness 360, an open-source toolkit to help developers detect and mitigate bias in AI models, and AI Explainability 360, a comprehensive open-source library of algorithms that support the explainability of machine learning models. By making these tools freely available, IBM has not only implemented its own principles but also empowered others to do the same.

Case Study 3: Microsoft's AI and ethics in engineering and research (AETHER) Committee

Six ethical principles—inclusivity, fairness, dependability and safety, privacy and security, transparency, and accountability—guide Microsoft's approach to responsible AI development and deployment. To operationalize these principles, Microsoft established the AETHER Committee, which provides guidance on AI-related ethical issues and conducts reviews of sensitive AI projects.

One example of AETHER's work is the decision to limit the use of Microsoft's facial recognition technology by law enforcement agencies due to concerns about potential misuse and infringement of individual rights. This decision reflects a commitment to the principles of fairness, privacy, and accountability.

These case studies indicate that responsible AI development and deployment are not only feasible but also beneficial. By adhering to ethical principles, engaging with stakeholders, and assessing long-term effects, companies like Google, IBM, and Microsoft have been able to navigate the complex AI landscape, make informed decisions, and contribute to the development of AI technology that is not just powerful but also trustworthy and beneficial for all.

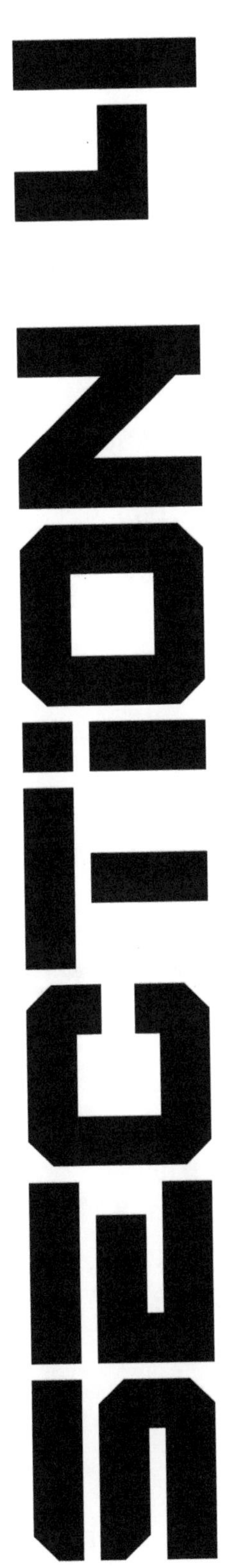

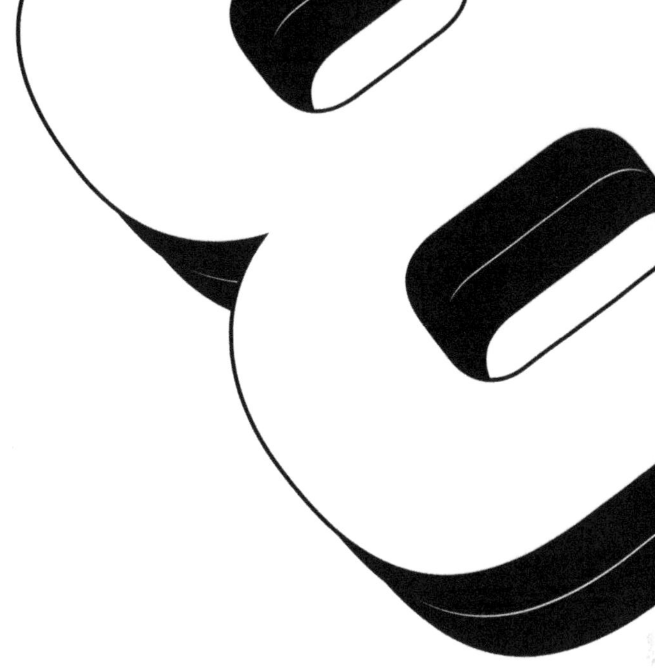

Practical Tools for Responsible AI

As the world continues to grapple with the ethical complexities of artificial intelligence (AI), an array of practical tools has emerged to aid developers in creating more responsible and ethical AI systems. These tools span various aspects of AI development, from bias detection and mitigation to explainability and transparency to privacy preservation.

Bias detection and mitigation

Bias can creep into AI systems through various channels, often via the training data. Several tools now exist to help identify and mitigate such biases. IBM's AI Fairness 360 is a comprehensive open-source toolkit that provides metrics to evaluate for fairness and algorithms to mitigate bias in models and datasets. It is an interactive experience that brings the concepts of fairness, bias, and mitigation strategies to the forefront.

Explainability and transparency

Artificial intelligence transparency involves understanding how AI systems make decisions. Google's What-If Tool provides a simple interface for developers to visually probe the behavior of their machine-learning models, encouraging a comprehensive grasp of the underlying processes.

IBM's AI Explainability 360 is an open-source library that supports the interpretability and explainability of data and machine learning models, allowing developers to choose from an array of algorithms to explain how their models are working.

Privacy preservation

Preserving privacy in the era of AI is a growing concern. OpenMined is an open-source community focused on researching, developing, and promoting tools for secure, privacy-preserving, value-aligned artificial intelligence. OpenMined uses technologies like federated learning, differential privacy, and encrypted computation to ensure the highest level of data privacy.

Trust and ethics

Microsoft's InterpretML is an open-source Python package that incorporates state-of-the-art machine learning interpretability techniques to help developers understand their models better, build trust, and adhere to ethical guidelines.

Regulatory compliance

Google's AI Hub provides enterprise-grade sharing capabilities, including end-to-end AI pipelines and out-of-the-box algorithms that comply with regulatory norms. It allows developers to easily share, collaborate, and reuse different elements of AI development.

These practical tools are not an exhaustive list but represent some exciting developments in the field of responsible AI. They demonstrate the tangible ways in which ethical considerations can be incorporated into the AI development process, providing a robust and practical foundation for developers to build AI systems that are fair, explainable, privacy-preserving, trustworthy, and compliant with regulations. With these tools in hand, the promise of responsible AI appears not just workable but eminently achievable.

CHAPTER NINE

The Exponential Evolution of AI: Predicting and Preparing for a Future Beyond Our Imaginations

On a distant horizon, the future of artificial intelligence unfolds—a vista so immense and complex that it defies our innate human capacity to perceive linear progression. The key to unlocking this understanding is a concept we often grapple with, and it is vital to our comprehension of the evolution of AI—exponential growth.

Exponential growth is an alien concept to our instinctual understanding of the world. We are, by nature, linear thinkers. We comprehend time and progression in a direct, one-dimensional way—one foot in front of the other, one page after another. But the evolution of AI does not conform to this linear path. It expands and evolves in a burst of exponential growth, where every step forward is not just an addition but a multiplication. It is difficult for us to grasp the idea of something doubling, then quadrupling, and eventually becoming sixteen times more potent.

This is not a failing of our cognition but a result of our evolutionary past. Our ancestors, who wandered the Savannahs of Africa, did not need to understand exponential growth. They survived and thrived based on linear, direct relationships—if you run twice as fast, you will catch twice as much prey; if you gather twice as long, you will have twice as much food. These linear equations were the lifeblood of our survival, and so they became the bedrock of our instinctual understanding of the world.

Yet here we stand on the precipice of an era defined by exponential growth. From Moore's Law, which postulated the doubling of computer processing power every two years, to the increasingly rapid advances we see in AI, exponential growth is the beating heart of our technological age. It is the reason why AI, once a curious novelty, is now an omnipresent force reshaping our world.

Indeed, the concept of exponential growth is so deeply interwoven into AI's evolution that one cannot be understood without the other. The objective of this chapter is to examine the intricate relationship between AI and exponential growth and analyze what this shows for our future.

We stand at the dawn of an era where AI, propelled by the relentless engine of exponential growth, has the potential to surpass our wildest imaginations. The path ahead is uncharted and full of uncertainties, but it is our responsibility to predict and prepare for this future, no matter how unfathomable it may seem.

This journey of understanding starts with a simple yet profound truth: the future of AI will not be a straight line but an exponential curve, reaching heights we can barely conceive. And while this concept may be difficult to grasp, it is essential for navigating the vast potential and pitfalls that the future of AI promises. Let us start this journey by leaving the linear path behind and embracing the exponential climb ahead.

SECTION ONE

The Mathematics of Exponential Growth and AI

To fully appreciate the trajectory of AI's evolution, we must first understand the mathematics underlying it. Exponential growth, often visualized as an upward-curving line on a graph, is more than just a numerical concept; it is a key driver of AI's breathtaking ascent.

Exponential growth deals with the principle of doubling. A quantity subject to exponential growth does not increase by a consistent amount over regular intervals; instead, it doubles. This might seem like an abstract concept, so let us ground it in a simple yet revealing example.

Consider a water lily growing in a pond. Suppose that this water lily doubles in size each day. On the first day, it covered a small area of the pond. By the second day, it had covered twice that area. On the third day, it covers four times the area, and so forth. After thirty days, it completely covers the pond.

Now, here is the catch: when do you think the pond is half covered? Many would intuitively say the fifteenth day, but the answer is the twenty-ninth day. This is the essence of exponential growth—the sudden, seemingly out-of-nowhere explosion of growth that characterizes the "hockey stick" curve we associate with exponential processes.

This principle is deeply embedded in the development of AI. It is what makes AI seem to come out of nowhere, suddenly transforming from a fringe technology to a dominant force in our world. When applied to AI, it is not just the size of a water lily we are dealing with—it's the processing power, the sophistication of algorithms, and the volume of data we can analyze.

Moore's law supports this phenomenon, stating that the number of transistors on an integrated circuit (and therefore processing power) doubles every two years. It has propelled the digital revolution and laid the groundwork for the advent and exponential growth of AI.

Another example is the development of AI algorithms. Each new iteration of an AI algorithm does not just add a little more capability—it often doubles, triples, or even quadruples the power or efficiency of the last version. Therefore, we've seen AI leap from being able to beat humans at checkers, chess, and Go to complex video games like Dota 2 and StarCraft II in a relatively short span of time.

The implications of exponential growth in AI are profound. This is not a linear progression where we can see the path ahead clearly. It is a curve that rapidly shoots upward, making the future increasingly difficult to predict. Yet understanding this principle is key to both anticipating the trajectory of AI and managing its impact on our world. The pond is filling rapidly, and understanding the nature of the water lily's growth is the first step in preparing for the world when it is fully covered.

SECTION TWO

AI's Exponential Trajectory: Cases and Data

Artificial intelligence's journey so far illustrates a classic case of exponential growth. This rapid expansion, often concealed until it reaches a tipping point, is visible in various dimensions of AI—from its computational power to its performance capabilities and its permeation into various sectors.

Take, for instance, the evolution of AI's computational power. According to a 2018 report by OpenAI, the computational power used in the largest AI training runs has been doubling every 3.4 months since 2012, outpacing Moore's Law by a significant margin. This exponential increase in computational power is not merely a number game; it has concrete implications for the kinds of tasks AI can execute and the efficiency with which it can perform them.

The effects of this computational escalation are evident in the AI's performance. Models like GPT-3 serve as examples of AI's capacity to comprehend and produce human language. GPT-3, with its 175 billion machine learning parameters, is a quantum leap from its predecessor, GPT-2, which had 1.5 billion parameters. In comparison, GPT-4 has 1.7 trillion parameters, a further exponential growth from GPT-3. GPT-4 has become exponentially more complex, enabling it to produce text that is frequently indistinguishable from human-written text—a feat that was unthinkable a few years ago.

Another interesting illustration of AI's exponential trajectory is its deepening penetration in various sectors. AI applications have swiftly moved from being predominantly used in tech-centric sectors to becoming ubiquitous across industries, including health care, education, finance, and even the arts. AI now aids in diagnosing diseases, personalizing learning, predicting financial markets, and creating artwork—tasks that were previously regarded to be the exclusive domain of human intelligence.

Consider the healthcare sector, where AI has made remarkable strides. In 2017, a Stanford study showed that an AI algorithm could identify skin cancer with the same accuracy as board-certified dermatologists. Furthermore, in 2021, Google's DeepMind AI achieved a significant milestone by developing an advanced system capable of predicting the 3D structures of proteins, effectively solving a longstanding and complex challenge in the field of biology that had persisted for over fifty years. Currently, DeepMind has predicted nearly every known protein, opening up a myriad of possibilities for advancements in medicine, biotechnology, and scientific research.

However, as we marvel at the accomplishments of AI, it is crucial to remember that exponential growth also magnifies challenges. The speed of AI's development makes it difficult for regulations, ethical guidelines, and societal understanding to keep pace. Each leap in AI capabilities brings forth new questions about privacy, security, job displacement, and the very nature of human identity in an increasingly automated world.

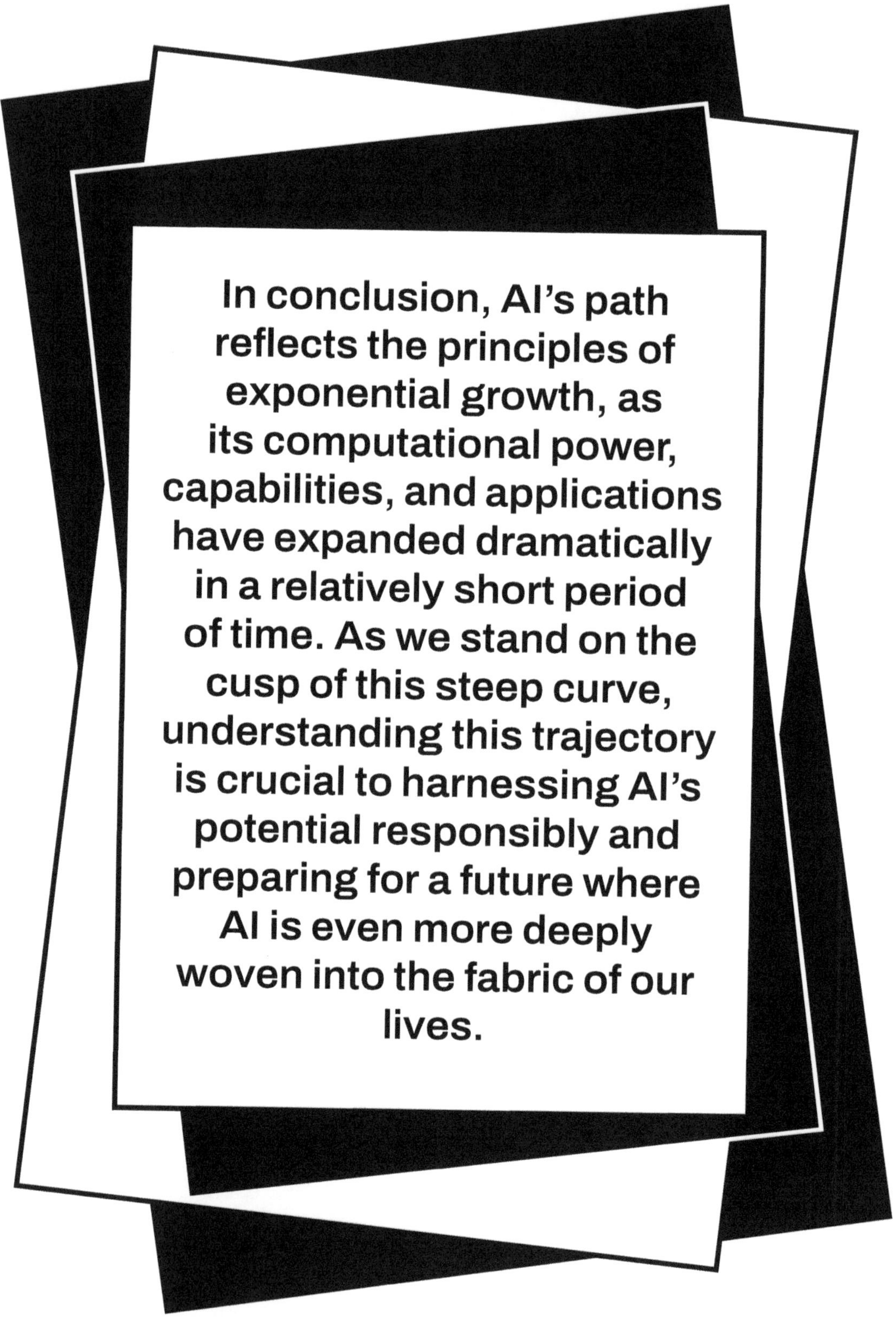

In conclusion, AI's path reflects the principles of exponential growth, as its computational power, capabilities, and applications have expanded dramatically in a relatively short period of time. As we stand on the cusp of this steep curve, understanding this trajectory is crucial to harnessing AI's potential responsibly and preparing for a future where AI is even more deeply woven into the fabric of our lives.

SECTION THREE

Linear Versus Exponential: A Contrast in AI's Growth Trajectory

Understanding the nature of AI's growth trajectory involves a clear contrast between linear and exponential growth. Human intuition, largely grounded in linear perception, struggles to comprehend the staggering implications of exponential growth. This disparity becomes starkly evident when we apply these concepts to the development and impact of AI technologies.

In a linear growth scenario, progress is steady and incremental. If you were to plot this on a graph, you would see a straight line moving predictably from one point to the next. For instance, consider a situation where you take thirty steps linearly, with each step advancing you one meter. After thirty steps, you would be thirty meters away from your starting point. This progression is intuitive and easy for us to visualize because it is directly proportional: double the steps, and you double the distance.

Now, let us consider an exponential growth scenario, where instead of taking thirty linear steps, you take thirty exponential steps. In this case, each step is twice the previous one. The first step takes you one meter, the second two meters, the fourth eight meters, and so on. By the thirtieth step, you would have astonishingly circumnavigated the earth approximately twenty-six times! This vivid example illustrates the vast difference between linear and exponential growth and why our linearly wired brains struggle to grasp the latter's implications.

So how does this relate to AI? When we think about AI's evolution, we are not dealing with a linear progression. Instead, we are grappling with an exponential curve. This is clear when we look at the leaps in computational power, performance efficiencies, cost-effectiveness, and the rapid diffusion of AI across various sectors.

Consider the evolution of machine learning models. In the early 2010s, models with a few million parameters were considered state-of-the-art. Fast-forward to 2023, and OpenAI's GPT-4 model boasts 1.7 trillion parameters—an incredible exponential increase in just a decade. What is more, the capabilities of these models have grown in parallel. AI systems now routinely perform tasks like language translation, image recognition, and game-playing, which were once considered monumental challenges.

This exponential growth, however, does not just bring about astonishing advancements. It also accentuates the challenges and risks associated with AI. Issues around privacy, security, bias, and job displacement become progressively more complex and pressing as AI evolves at an exponential pace. Ethical and regulatory frameworks, which typically evolve linearly, struggle to keep up with the pace of AI development, leading to a widening gap that needs urgent attention.

Contrasting linear and exponential growth helps us understand not only the spectacular progress of AI but also the urgency to address the challenges it brings. Understanding the exponential trajectory of AI is crucial to harnessing its benefits while reducing its risks.

SECTION FOUR

The Human Brain and Exponential Thinking

The irony of exploring the exponential evolution of AI is that it originates from human creativity, but we cannot fully comprehend it. The simple reason is that our brains, remarkable as they are, have not been naturally wired to think exponentially. They developed in a world where survival depended on understanding linear, cause-and-effect relationships rather than exponential ones.

This cognitive bias toward linear thinking has deep evolutionary roots. Early humans needed to estimate distances, quantities, and time intervals accurately for survival, but they did not need to compute compound interest or predict viral transmission rates. Hence, our brains developed heuristics, or mental shortcuts, to deal with the kind of linear problems our ancestors faced. This perspective has been studied in cognitive psychology, particularly in the work of Daniel Kahneman and Amos Tversky, who showed how these mental shortcuts often lead to systematic errors in prediction or cognitive biases.

One such cognitive bias that hinders our understanding of exponential growth is the "anchoring effect." It denotes our propensity to heavily rely on the initial piece of information we receive (the "anchor") when making decisions. Our incremental experience with technology makes it hard to adjust our expectations for AI's rapid growth.

Another relevant cognitive bias is the "linear assumption bias," where we naturally assume that the future will be like the past and that changes will occur in a steady, linear manner. This bias makes it hard for us to accept that AI could transform our world in ways that may seem unimaginable based on our past experiences.

The "availability heuristic" is another cognitive shortcut that affects our ability to predict exponential growth. This heuristic leads us to base our predictions on information that is readily available to us rather than seeking all necessary and relevant data. Given that the most readily available information typically pertains to the recent past, we underestimate the potential for rapid, exponential change.

Overcoming these cognitive biases is a significant challenge, but it is not insurmountable. It requires deliberate effort, education, and practice. We must train ourselves to think in terms of rates of change rather than absolute quantities and to appreciate the power of compounding. We need to shift from a static, fixed mindset to a dynamic, growth-oriented one. This cognitive shift is crucial not only for those directly involved in AI development but for all of us as we navigate an increasingly AI-driven world.

Understanding our cognitive biases concerning linear thinking is the primary step toward appreciating the exponential growth of AI. As we develop strategies to overcome these biases, we will be better equipped to anticipate and prepare for a future where AI could redefine the realm of possibility.

Our cognitive biases and AI's future trajectory

Our cognitive biases

As we look further into the intricacies of AI's growth, it becomes evident that our cognitive biases can function as barriers, distorting our perception of AI's future trajectory and its implications. This misperception has ramifications not just for technologists but also for policymakers, business leaders, educators, and indeed, society at large. Let us further explore these possible obstacles.

The "linear assumption bias" could lead us to drastically underestimate the pace at which AI technologies might evolve and permeate our lives. Experts like Ray Kurzweil posit that if we assume AI's future development will proceed at the same pace as in the past, we might fail to anticipate a future only a few years away where AI could outperform humans in the most economically valuable work. Such underestimation could lead to a lack of preparation and adaptation, causing significant socioeconomic disruptions.

Our cognitive biases … trajectory

The "anchoring effect" might make us slow to revise our beliefs about AI's potential by considering new data. This bias could cause a lack of recognition of AI's transformative potential or an underestimation of its associated risks. For instance, if we are anchored to the idea that AI is merely a tool for automating mundane tasks, we might overlook its potential to create new kinds of jobs, industries, and societal structures or its capacity to cause unprecedented shifts in power dynamics.

The "availability heuristic" could skew our perception of AI's implications based on sensationalist headlines or dystopian science fiction rather than rigorous scientific evidence. While immediate concerns like algorithmic bias, data privacy, or AI-enabled surveillance may be downplayed, misplaced fears may arise from an overemphasis on the risk of malevolent, super-intelligent AI.

Understanding these cognitive biases and how they can hinder our comprehension of AI's future trajectory is crucial. Recognizing these mental stumbling blocks allows us to mitigate their effects, paving the way for a more nuanced and accurate understanding of AI. It encourages us to question our assumptions, seek diverse perspectives, and update our beliefs in response to new evidence.

This knowledge underscores the importance of inclusive and interdisciplinary dialogue in shaping AI's future. Technologists, ethicists, social scientists, policymakers, and citizens need to collaborate, bringing together diverse expertise and perspectives to address the hurdles and opportunities arising from AI.

Being aware of AI's exponential growth is not just a matter of interpreting the mathematics; it is also about recognizing and overcoming our cognitive biases. By doing so, we can foster a more informed and constructive discourse about AI, aligning its trajectory with our collective values and aspirations, and harnessing its potential to create a future where humanity flourishes.

Embracing exponential thinking: navigating the future of AI

The ability to think exponentially has evolved from an intellectual curiosity into a crucial skill in a world where AI is developing at an exponential rate. The key to unlocking this potential lies not only in understanding the mathematical principles of exponential growth but also in adopting strategies to mitigate cognitive biases and enhance our predictive accuracy.

Education and awareness form the foundation of this process. A comprehensive understanding of AI's capabilities, its trajectory, and its potential societal implications can help dispel misconceptions and align our expectations with reality. This includes the study of AI technology itself along with its broader ethical, legal, and societal dimensions. Institutions of learning, both traditional and non-traditional, have an important role in fostering this understanding and promoting a culture of perpetual learning and adaptability in the face of technological change.

We must also cultivate the habit of **questioning our assumptions and biases**. This can be achieved through practices like "red teaming," where a group of individuals critically examines an issue from all possible angles, or by leveraging AI itself, using it as a tool to identify and mitigate biases in our decision-making processes.

The third strategy involves **fostering diversity and interdisciplinary collaboration**. The future of AI is not just a matter for technologists; it involves a myriad of social, ethical, and economic dimensions. Engaging diverse perspectives—from ethicists, social scientists, policymakers, and end-users—can help uncover blind spots, challenge groupthink, and result in more robust predictions and solutions.

Next, **embracing uncertainty and adopting a scenario-based approach** can enhance our predictive accuracy. While it is impossible to predict the future with complete certainty, we can imagine a range of plausible scenarios based on current trends and uncertainties. These scenarios can serve as a "compass," guiding our strategic decisions, helping us prepare for different contingencies, and enabling us to respond more effectively as the future unfolds.

Finally, the strategy of **continuous learning and adaptation** is crucial. In a rapidly evolving field like AI, what worked yesterday might not work tomorrow. We must remain open to new data, be willing to revise our beliefs, and adapt our strategies accordingly. This requires humility, curiosity, and resilience—qualities that machines, for all their prowess, cannot replicate.

As a last remark, thinking exponentially about AI involves not just understanding mathematics but also recognizing and overcoming our cognitive biases, fostering diversity, embracing uncertainty, and adopting a mindset of continuous learning and adaptation. While these strategies do not guarantee perfect foresight, they increase our capacity to navigate the challenges and opportunities of the AI age, enabling us to steer its trajectory toward a future where both humanity and technology can thrive.

Envisioning the exponential future of AI: scenarios across spheres

As we stand on the precipice of the AI age, the vast spectrum of potential futures can be daunting. Nevertheless, it is essential to confront these possibilities, acknowledging the myriad ways AI's exponential growth could reforge our societies. Let us envision scenarios across three vital sectors: the economy, healthcare, and education.

1 The economy: AI as the catalyst of abundance

In this scenario, AI functions as an engine of unprecedented economic growth and prosperity. Automation of routine tasks liberates human labor from drudgery, freeing us to pursue more creative and fulfilling endeavors. Artificial intelligence systems augment human capabilities, driving productivity to new heights and creating wealth on a scale hitherto unimaginable.

Smart algorithms identify market inefficiencies, democratize access to financial services, and foster economic equality. Advanced predictive models enable businesses to optimize supply chains, reduce waste, and enhance sustainability. Governments harness AI for smart urban planning, resulting in cities that are more livable and resilient.

2 Health care: the dawn of precision medicine

In the health care domain, AI's exponential growth could usher in an era of precision medicine, where treatments are tailored to individual patients' genetic, environmental, and lifestyle factors. AI algorithms analyze vast troves of medical data, discerning subtle patterns that elude human doctors and enabling early detection of diseases like cancer or Alzheimer's.

Robot-assisted surgeries achieve levels of precision and consistency beyond human capability, reducing complications and improving patient outcomes. Virtual nursing assistants provide round-the-clock care, offering comfort and companionship to the elderly and chronically ill. AI-powered telemedicine democratizes access to health care services, reaching the most remote and underserved communities.

Education: personalized learning for all

In the sphere of education, AI could pave the way for truly personalized learning. AI tutors adapt to each student's unique learning style, pace, and interests, fostering a love of learning and cultivating each individual's full potential. Real-time analytics identify gaps in understanding, allowing for timely interventions and support.

Virtual reality, which uses AI to power it, immerses students in interactive, experiential learning environments, improving the effectiveness and engagement of education. Artificial intelligence tools democratize access to quality education, breaking down geographical, socioeconomic, and physical barriers.

These scenarios are not without their challenges. Each harbors potential pitfalls and ethical quandaries, from job displacement and inequality in the economic sphere to privacy concerns and health disparities in health care to digital divides and algorithmic biases in education. By envisioning these futures, we can begin to chart a course toward a future where AI serves as a tool of empowerment, prosperity, and human flourishing, guided by our shared values and aspirations.

SCENARIO 1:

the exponential economy: the transformation of wealth and work

The economic landscape stands at the brink of a seismic shift. With AI's exponential growth, the economy of tomorrow might be almost unrecognizable. In this new world order, AI influences not only the economy—it becomes a prime mover, sparking new industries, reshaping labor markets, and even nudging us toward a post-scarcity society.

Rise of new industries

Artificial intelligence is not just another industry. It is a foundational technology, akin to electricity or the internet, and its tentacles will reach into every sector of the economy. But beyond that, AI will birth entirely new industries. Consider the burgeoning field of AI ethics, where professionals guide the responsible development and deployment of AI systems, or the domain of AI law, where legal experts grapple with questions of liability, privacy, and intellectual property in the AI age.

In health care, we might see the rise of AI diagnostics companies that can detect diseases from medical imaging with superhuman accuracy. In education, AI tutoring services could offer personalized learning at scale. These are just a smattering of the myriad industries that could emerge from the crucible of AI's exponential growth.

Job displacement and the evolution of work

As AI automates routine tasks, certain jobs will inevitably become obsolete. This transition could be tumultuous, displacing workers and exacerbating social inequalities. History teaches us that technology-induced job displacement is not the end of the story. Just as the industrial revolution eventually led to a net increase in jobs, the AI revolution could engender a similar economic renaissance.

New jobs will be created—some in new industries, others in existing fields that require a human touch. The care economy, creative industries, and roles requiring complex decision-making or high emotional intelligence are likely to flourish. Moreover, AI will not just replace tasks but also augment human abilities, enabling us to work smarter and achieve more.

Toward a post-scarcity society

The most intense economic implication of AI could be its potential to usher us into a post-scarcity society. Artificial intelligence, in combination with other technologies like additive manufacturing and renewable energy, could drastically reduce the cost of goods and services. As AI systems become more capable, resources that were previously scarce could become abundant.

This transition would not be without its challenges. It would cause a radical rethink of our economic systems and societal structures. Concepts like universal basic income or universal basic services might move from the fringes to the mainstream. Even our notions of work and purpose might need to be reimagined in a world where material needs are met.

In sum, AI empowers the exponential economy, holding tremendous promise, but it is not a guaranteed utopia. It presents us with a set of choices. Will we harness AI to create broad-based prosperity, or will we allow it to exacerbate inequality? The answers to these questions lie not in the technology itself but in us—in the policies we enact, the ethical guidelines we establish, and the vision of the future we choose to pursue.

SCENARIO 2:

health 2.0: AI as the catalyst for a health care revolution

Imagine a world where each individual's health is not a matter of chance but a meticulously designed, personalized endeavor. A world where diseases are not just managed but preempted, where diagnostics are not only accurate but predictive. This is the promise of AI in health care—the potential dawn of health 2.0.

The dawn of personalized medicine

There is no one size does not fit all, especially when it comes to health. Artificial intelligence's ability to crunch vast amounts of data could herald a new era of personalized medicine. By analyzing everything from your genetic makeup to your lifestyle habits and even your social media posts, AI systems could create a holistic, personalized health profile.

These profiles could inform individualized treatment plans tailored to your unique physiology and circumstances. Cancer treatments, for instance, could be fine-tuned based on your genetic vulnerabilities. Mental health interventions could be crafted based on your cognitive styles and life experiences. Diet and exercise plans could be optimized for your metabolic characteristics.

Revolutionizing diagnostics

AI could revolutionize the diagnostic process, which is frequently a time-consuming game of guesswork. Machine learning algorithms have already demonstrated their prowess in interpreting medical imaging, often outperforming human experts. But AI's potential in diagnostics goes beyond just image analysis.

Consider liquid biopsies, a noninvasive procedure that detects cancer by looking for DNA fragments shed by tumor cells in the bloodstream. The challenge here is finding the proverbial needle in the haystack—the few cancerous DNA fragments among millions of healthy ones. This is where AI's pattern-recognition capabilities shine.

AI could make diagnostics predictive. By integrating and analyzing disparate data—genomic, phenotypic, and environmental—AI could identify disease risk factors and flag potential health issues before they manifest, shifting health care from a reactive to a proactive endeavor.

A leap in lifespan

Finally, AI's exponential growth might lead to a leap in our lifespan. This could happen through a combination of early disease detection, personalized interventions, and breakthroughs in understanding aging at a molecular level. The latter, in particular, is an area where AI's data-crunching prowess could help unearth insights that lead to longevity-boosting interventions.

In essence, health 2.0, underpinned by AI, could transform our healthcare system from a "sick care" model to a "truly health care" one. However, this rosy vision comes with caveats. Issues of data privacy, algorithmic bias, and accessibility must be addressed for us to fully reap the benefits of AI in health care. As with the exponential economy, the future of health is not just about technology but about the choices we make as a society.

SCENARIO 3:

education for everyone: the democratization of learning through AI

We are on the brink of an educational revolution. Barriers to learning are being dismantled, and opportunities are being extended to learners of all backgrounds and abilities. Through AI-driven adaptive platforms, educational content is becoming more inclusive, engaging, and accessible. This transformative movement marks a critical step towards a future where education truly becomes a fundamental right for everyone, driving societal progress and human development in an increasingly interconnected world.

Personalized learning: education tailored for the individual

Traditional education has been a game of averages. The designed curriculum and pedagogical methods are for the average student, who, in reality, doesn't exist. AI has the potential to disrupt this model by facilitating a shift towards personalized learning.

Artificial intelligence-based learning platforms could curate educational content tailored to each student's unique learning style, pace, and interests. Struggling with calculus? The AI could break down the concepts into simpler, digestible chunks and provide targeted exercises to help you master them. Interested in ancient history? The AI could guide you through a curated learning journey, replete with interactive multimedia content.

The implications are profound. By tailoring education to the individual, we could unlock human potential on a previously unimaginable scale. The slow learner, the gifted child, the late bloomer—each could have an educational pathway designed to help them flourish.

Global access to quality education: leveling the playing field

The democratizing potential of AI in education extends beyond personalization. By leveraging AI and digital technology, we could make quality education accessible to all, regardless of geography, socioeconomic status, or physical ability.

Imagine an AI-powered online platform that provides access to world-class curriculum and advanced pedagogy. A child in a remote village in Africa or a disadvantaged neighborhood in Detroit could learn from the best educators in the world. AI-powered translation and transcription tools could overcome language barriers, while AI tutors could provide on-demand support, ensuring no student is left behind.

This democratization of education could be a powerful equalizer, breaking down barriers of privilege and geography. But it is not without challenges. Ensuring digital access for all, preventing the digital divide from becoming a learning divide, and maintaining the human touch in education are issues that need addressing.

It is important to remember that we hold the key to unlocking AI's full potential in education as we move forward. We need to guide this technological revolution with a sense of responsibility and a commitment to inclusivity, always remembering that education is about nurturing the human spirit.

SCENARIO 4:

ethical quandaries: navigating the moral labyrinth of advanced AI

As we look toward the future of AI, we see a landscape full of ethical challenges quickly approaching. From privacy concerns to algorithmic bias and the potential misuse of AI, these dilemmas force us to grapple with questions that strike at the very core of our societal values.

The prism of privacy in a hyperconnected world

With AI dominating and data being the new oil, privacy will become a top concern in the future. AI systems, driven by a voracious appetite for data, will permeate every facet of our lives, constantly collecting, analyzing, and learning from our digital footprints.

In such a scenario, what happens to the sanctity of our private lives? What safeguards are in place to ensure that our personal information is not misused? These questions underscore the urgent need to establish robust data governance frameworks and ensure transparency in how AI systems collect, store, and use data.

Algorithmic bias: the shadow of our past

AI systems are mirrors, reflecting the biases present in the data they are trained on. They can perpetuate and amplify historical patterns of discrimination, leading to unfair outcomes. For instance, an AI-driven hiring system trained on past data could inadvertently discriminate against certain demographic groups.

Overcoming this challenge requires concerted efforts to improve data representativeness and develop algorithms that are robust to bias. It also underscores the need for diverse teams in AI development to ensure a plurality of perspectives and minimize unconscious bias.

The double-edged sword of AI

Like any technology and just like the human mind, AI is a tool that can be used for good or bad. An AI that can predict health risks can save lives, however in the wrong hands, it can also be a tool for surveillance and control. This potential for misuse, especially as AI systems become more powerful, poses a significant ethical challenge.

Navigating this moral maze requires an initiative-taking approach. We need to develop strong regulatory frameworks, foster international cooperation, and ensure that the development and deployment of AI align with our societal values.

As we journey into this brave new world, we are not just passive observers. We are the architects of this future. By engaging with these ethical quandaries head-on, we have the opportunity to guide the evolution of AI in a way that reflects our collective aspirations for a fair, inclusive, and fair society.

SCENARIO 5:

democracy 2.0: AI as a catalyst for political transformation

AI's potential to transform the political landscape is as profound as it is complex. In this scenario, we will explore how AI could revolutionize democracy, potentially leading to a new form of government, improved policies, and increased individual benefits, while also addressing the dangers inherent in such a transformation.

Artificial intelligence has the potential to be a catalyst for what we may term "Democracy 2.0," a new form of participatory governance. First, AI can enhance the decision-making process by offering data-driven insights, predictive modeling, and real-time analytics. Governments could make more informed choices by using AI to predict the outcomes of different policy decisions based on vast amounts of data. This could lead to more efficient resource allocation, improved public services, and a heightened ability to respond to societal challenges.

Consider the use of AI in formulating public health policy. By analyzing data from a wide range of sources, AI could help identify trends and patterns that human analysts might miss, thereby enabling policymakers to devise strategies that are more responsive to public health needs.

Artificial intelligence also holds the promise of democratizing the policymaking process itself. By leveraging AI-powered platforms, governments could potentially garner citizen input on a massive scale, leading to more participatory decision-making. Citizens could voice their opinions, vote on issues, and contribute to policy development in real-time, fostering a form of direct democracy that's significantly more inclusive and responsive than traditional models.

However, the potential benefits of AI in democracy do not come without significant risks. There is the obvious concern about privacy. With governments having access to vast amounts of data, the potential for misuse or mishandling of information is a valid concern that must be addressed with robust data protection laws and regulations.

Moreover, the use of AI in policy decision-making could inadvertently result in the marginalization of certain groups. Algorithmic bias is a well-documented issue in AI, and if left unchecked, it could perpetuate existing social inequities. For instance, if an AI system was trained on data that reflected biased policing practices, it might recommend policies that further entrench these biases.

The potential for AI to be used as a tool for surveillance and control by authoritarian regimes cannot be overlooked. Without proper checks and balances, AI could be used to suppress dissent, manipulate public opinion, or infringe upon civil liberties.

With a potential future in mind, we must approach the integration of AI into our democratic processes with both optimism and caution. The promise of a more efficient, inclusive, and responsive democracy is tantalizing, but we must remain vigilant to the risks.

Striking a balance between leveraging AI's potential and safeguarding democratic values will require an ongoing, nuanced conversation among technologists, policymakers, and citizens. As we look to the future, let us strive to harness the power of AI not just to reimagine our democratic systems but to reinforce the very principles upon which they stand: liberty, equality, and justice for all.

SECTION FIVE

Navigating the Exponential Landscape: A Guide to Thriving in an AI-Driven Future

As we stand on the brink of this exponential leap in AI, it is important that we take the necessary steps to prepare for the significant changes it will bring to our lives, businesses, and communities. Yet preparing for such a future is not merely about bracing for impact but about actively shaping and navigating this landscape.

Individuals: lifelong learners in an AI world

For individuals, the rise of AI causes a reorientation toward lifelong learning. The exponential rate of AI development means that skills will become obsolete at an unprecedented pace. As AI automates routine tasks, the jobs of the future will place a premium on creativity, emotional intelligence, and adaptability—skills uniquely human and difficult for AI to replicate.

To stay relevant in this shifting landscape, individuals must become comfortable with continually learning new skills and adapting to changing circumstances. In this era of AI, curiosity and adaptability are not merely virtues; they are survival skills.

Corporations: innovate, adapt, or perish

For corporations, the rise of AI presents both a challenge and an opportunity. On the one hand, it threatens to disrupt traditional business models and render obsolete those companies that fail to adapt. On the other hand, it offers opportunities for tremendous productivity gains and the creation of innovative products and services.

To navigate this landscape, companies need to actively invest in AI capabilities. This includes not only technological investments but also investing in people and processes—fostering a culture that embraces change, encouraging cross-disciplinary collaborations, and promoting ethical AI use.

Societies: the urgent need for inclusive policies

For societies, the exponential growth of AI poses questions about inequality, privacy, and governance. As AI permeates every aspect of life, there is a danger that the benefits of AI will accrue only to those who control the technology, exacerbating societal inequalities.

To prevent this, there is a need for inclusive policies that ensure broad access to AI's benefits. This includes investing in public education to equip people with the skills required for an AI-driven future, establishing robust privacy protections, and developing regulatory frameworks to safeguard the responsible application of AI.

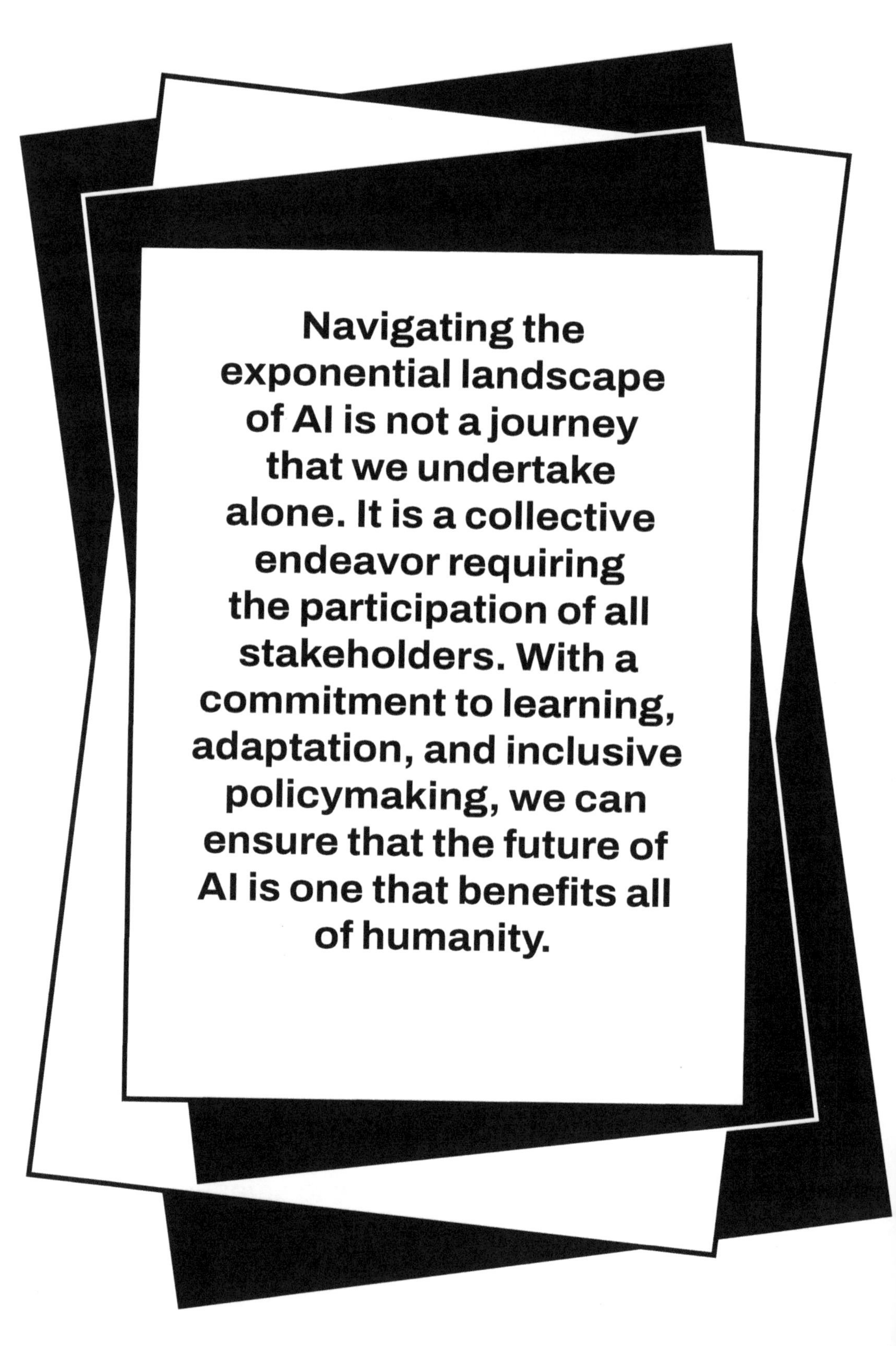
Navigating the exponential landscape of AI is not a journey that we undertake alone. It is a collective endeavor requiring the participation of all stakeholders. With a commitment to learning, adaptation, and inclusive policymaking, we can ensure that the future of AI is one that benefits all of humanity.

SECTION SIX

Ensuring Policy, Regulation, and Ethics Match AI's Pace

In the swiftly advancing field of artificial intelligence, policymaking, regulations, and ethical considerations are vital. This is the compass guiding us through the unfolding terrain, setting the boundaries within which AI operates, and ensuring that its power is harnessed for the greater good.

Policymaking: a guiding beacon

Policymaking in the context of AI is akin to creating a blueprint for a city that has not yet been built. In this rapidly growing landscape, policy must both anticipate and react to changes, providing a flexible framework that safeguards society without stifling innovation. It is a delicate balance to strike, requiring an intimate understanding of the technology and its implications.

The stakes are high. Artificial intelligence has the potential to reshape economies, redefine work, and rewire social structures. Effective policymaking ensures that this transformation is steered toward fair outcomes, distributing benefits broadly, and mitigating potential harms.

Regulation: guardrails in the AI landscape

Regulations serve as the guardrails in our exponential journey, preventing missteps that could lead to harmful outcomes. They provide necessary constraints on AI's use, tackling concerns such as data privacy, algorithmic fairness, and accountability.

However, given the pace of AI's development, regulations risk being reactive and falling behind technological advancements. To counter this, regulators must adopt a forward-looking stance, crafting flexible regulations that can adapt to unforeseen developments.

International cooperation is essential. Given the global nature of AI, a fragmented regulatory landscape could lead to "regulatory arbitrage," where AI development moves to regions with the laxest regulations, potentially endangering global norms and standards.

Ethics: the compass of AI development

Ethics serves as the compass of AI development, providing the moral principles that guide decision-making. As we explore unexplored territories, ethical considerations help us answer complex questions, such as, who benefits from AI? Who bears the risk? How do we ensure fairness and justice?

Given the societal implications of AI, it is not enough for it to be technically sound; it must also be ethically justifiable. Ethical AI requires a commitment to principles such as respect for human rights, transparency, and fairness. It demands that we consider not just what we can do with AI but what we should do with it.

In conclusion, as we stand at the edge of an AI-driven future, it is imperative that policymaking, regulations, and ethics keep pace with AI's growth. Only then can we ensure that the power of AI is harnessed responsibly and its benefits are shared equally.

ARTIFICIAL

CON
CLU
SION

INTELLIGENCE

As this book comes to an end, let us contemplate the profound journey we've taken in exploring AI's exponential growth, its implications, and the strategies required to navigate this new terrain.

We began our expedition by grappling with the mathematical principles of exponential growth and how they manifest in the realm of AI. It is a concept that challenges our instinctual understanding of growth, typically predicated on linearity, but it's one that is central to appreciating the pace and scale of AI's evolution.

We acknowledged our cognitive biases and the human propensity toward linear thinking. We saw how these biases could impede our understanding of AI's potential trajectory and its transformative implications. However, we also discussed strategies to enhance our ability to think exponentially, an essential skill in an increasingly AI-driven world.

Next, we dared to envision future scenarios born out of AI's exponential growth, each scenario casting light on a different facet of society—the economy, health, education, and ethical quandaries. From the possibility of a post-scarcity society to personalized medicine, democratized education, and ethical challenges, we explored a wide range of potential outcomes. While these scenarios are speculative, they serve as valuable thought exercises, prompting us to consider the breadth and depth of change that AI could introduce.

Our journey through this chapter has been one of understanding, anticipation, and preparation. It has underscored the importance of developing robust policies, flexible regulations, and principled ethics to keep pace with AI's exponential growth. These elements form the scaffold upon which a fair and safe AI-driven society can be built.

In conclusion, this book has sought to explain the imperatives of understanding and preparing for AI's exponential growth. We are on an uncharted and uncertain path toward an AI-powered future, but with an understanding of exponential growth, thoughtful anticipation of potential outcomes, and the strategic use of policy, regulation, and ethics as guiding tools, we can navigate a future where AI is a force for widespread prosperity and human flourishing. An era is approaching where AI will be our progress partner, and we must ensure that we guide this partnership with wisdom, foresight, and an unwavering commitment to human values.

EPI LOGUE

THE DAWN OF A NEW ERA

We are at the edge of a new era, feeling both apprehension and excitement about the potential of artificial intelligence. We are custodians of a technology that has the power to redefine the contours of our civilization, a responsibility as immense as it is inspiring.

Throughout this chapter, we have examined exponential growth's intricate mechanics, artificial intelligence's transformative potential, and the challenges and opportunities that lie ahead. We have peered into the realm of possibilities, the prospects of an exponentially thriving economy, a revolutionized health care system, democratized education, and a society that grapples with new ethical considerations. These are not mere dreams or science fiction; they are potential realities on the horizon of our collective future.

The key to unlocking this future lies not just in our technical prowess but in our collective wisdom, our shared values, and our ability to harness AI's power for the greater good of humanity. We stand at a junction where we can choose to be passive observers or we can choose to be active participants, guiding the course of AI's development in ways that align with our deepest held values and highest aspirations.

We need to remember that technology, in all its forms, is but a tool—a reflection of our intentions. Therefore, we, through our decisions and actions, will determine the future of AI rather than algorithms or machines. We have the chance to shape artificial intelligence into a tool that can tackle our greatest challenges, unlock new opportunities, and help us aspire toward a future where everyone can thrive.

As we move forward, let us carry with us the understanding that with great power comes great responsibility. This is not just a motto for the world of superhero lore; it is a mantra for our times, a call to action for each one of us. It is a recognition of our collective responsibility to guide AI's growth for the betterment of all humanity.

In the end, the journey toward an AI-driven future is not just about the technology. It is about us—our vision, our courage, our compassion, and our collective will. It is about the kind of world we want to create and the steps we're willing to take to get there.

This is our moment to shape the future we want. Let us seize it with both hands. We have the opportunity to create a world where AI is a tool for prosperity, a beacon of progress, and a partner for a brighter tomorrow.

To the dawn of this new era, I say, let us step forward with hope in our hearts and a clear vision in our minds, ever reminded that our greatest ally in this journey is not just the machine but the indomitable spirit of human ingenuity.

GLOSSARY
GLOSSARY
GLOSSARY

adversarial AI.
Artificial intelligence techniques are used to fool other AI systems, often by providing misleading input to alter the output in a malicious way.

artificial intelligence (AI).
Computer science's branch that emphasizes the development of intelligent machines that can simulate human actions and responses, such as speech recognition software that transforms spoken words into written text.

artificial general intelligence (AGI).
A form of artificial intelligence capable of comprehending, acquiring, and employing knowledge across diverse tasks at a human level or beyond. It encompasses an AI system that demonstrates general problem-solving abilities akin to human intelligence.

big data.
Vast datasets that can be computationally analyzed to unveil patterns, trends, and correlations, particularly concerning human behavior and interactions. These datasets are typically of such immense size that traditional data processing methods are inadequate, necessitating advanced computational techniques for their analysis.

cognitive enhancement.
Enhancing the information processing systems of the mind and extending cognitive abilities, such as memory and attention. It seeks to improve cognitive functions through various means, including training, education, and technological interventions. The objective is to optimize cognitive performance and empower individuals to reach their full cognitive potential.

common sense.
The fundamental capacity to perceive, comprehend, and make judgments about the world, expected to be universally shared among individuals.

convergent instrumental goals.
The idea in AI safety is that several different AI systems with different ultimate goals could all have similar "intermediate" goals, such as self-preservation and resource acquisition.

deep learning.
Specific area within machine learning that utilizes artificial neural networks with numerous layers (deep architectures) to interpret and unravel complex patterns present in datasets.

embodiment.
In AI, this refers to robots or other physical systems that interact with the physical world, in contrast to AI systems that exist only in software.

existential risk.
A risk that threatens the entire future of humanity, such as a global catastrophic event that would annihilate intelligent life or permanently and drastically curtail its potential.

exponential.
An escalating growth rate that accelerates in proportion to the expanding overall number or size.

Fermi's paradox.
Apparent contradiction between the high probabilities of the existence of extraterrestrial civilizations and the absence of any evidence or communication with such civilizations.

friendly AI.
An AI that is designed to act in the best interests of humanity.

human-level AI
An artificial intelligence capable of executing any cognitive task achievable by a human being.

law of accelerating returns.
A theory proposed by Ray Kurzweil that states technology advances exponentially, rather than at a linear rate, as each development builds upon previous ones.

machine consciousness.
A field of research that aims to implement consciousness in artificial intelligence.

mind uploading.
The hypothetical futuristic process of scanning the physical structure of the brain accurately enough to create an emulation of the mental state and copying it to a computer in digital form.

Moore's law.
The observation that the number of transistors on a microchip doubles approximately every two years leads to an exponential increase in computing power.

neural network.
A computational system loosely inspired by the intricate connections found in animal brains, designed to learn and perform tasks by analyzing examples, often without explicit task-specific instructions.

optimization.
In the context of AI, this refers to the process of adjusting a system to improve its efficiency or effectiveness in solving a task.

oracle AI.
A theoretical type of artificial intelligence that does nothing except answer questions as accurately as possible and does not act in the world in any other way.

paperclip maximizer.
A thought experiment that illustrates the possible risks of an AI running amok. It involves an AI tasked with making as many paperclips as possible, which could ultimately result in the destruction of humanity if left unchecked.

quantum computer.
A specialized computer that harnesses the principles of quantum mechanics to execute specific types of computations with greater efficiency compared to traditional computers.

recursive self-improvement.
The ability of an AI system to iteratively improve itself, i.e., to make modifications to its own structure and functionality that increase its intelligence.

reinforcement learning.
A distinct category of machine learning in which an agent acquires the ability to interact with its environment by taking specific actions and then analyzing the outcomes or rewards associated with those actions.

reward function.
In reinforcement learning, a function that returns a scalar value that the learning agent aims to maximize through its actions.

seed AI.
A hypothetical AI system that is capable of recursive self-improvement. Starting with a basic system, it could redesign itself repeatedly, each iteration increasing its intelligence.

superintelligence.
A theoretical entity endowed with intelligence vastly exceeding the cognitive capabilities of even the most brilliant human intellects.

technological singularity. A theoretical point in time when artificial intelligence will have progressed to the point of greater-than-human intelligence, radically changing civilization.

transhumanism. An intellectual and philosophical movement that advocates for the use of technology to enhance the human condition, including enhancing intellectual and physical capacities and potentially achieving immortality.

Turing machine. A fundamental concept in computer science, serving as a mathematical model of computation that involves an abstract machine manipulating symbols on a tape based on a set of predefined rules.

Turing test. Evaluates a machine's capacity to exhibit intelligent behavior on par with or indistinguishable from that of a human.

universal artificial intelligence (UAI). Refers to an AI system with the ability to accomplish any task that a human can, essentially a more formal term for AGI.

vicarious embodiment. The application of an AI's capability or consciousness into a body that is not its own, allowing it to experience different situations and environments.

whole brain emulation (WBE). The hypothetical process of scanning and mapping the mental state (including long-term memory and self) of a particular brain and copying it to a computer.

zombie AI. A hypothetical AI that exhibits behavior indistinguishable from that of a human but without any form of consciousness or understanding.

our vision

In the present digital age, an era characterized by immediate access to a boundless array of information, we find ourselves in an ocean of opinions. While this wealth of data may satiate our curiosity on a superficial level, it often fails to provide the depth of understanding we crave—the kind of fundamental understanding that gives birth to a conscientious and comprehensive perception of the world. This is precisely where Mekiki Magazine comes into play. We stand as a beacon in the information storm, offering not just information but knowledge, wisdom, and insight.

At Mekiki Magazine, we believe in intertwining cutting-edge technology with the wisdom of industry experts to curate highly tailored content. With the integration of our unique proprietary content, we are revolutionizing how knowledge is disseminated, making it much more accessible globally. Our innovative approach harnesses technology to deliver insights that cater to global preferences and needs. We are committed to empowering our writers with advanced intelligence capabilities, both generative and predictive, to enhance their understanding of the interests of our readers. This, in turn, allows them to deliver content that truly resonates with their audience, providing unparalleled depth and value. At Mekiki Magazine, we are not just about informing; we are about transforming understanding, one reader at a time.

Our mission is to break down the walls between specialists and the broader audience, making all topics accessible and engaging to all. Each of our publications, whether a magazine, book, or online article, acts as a unique portal into realms of knowledge. We are not simply content creators but translators of intricate subjects, transforming abstract ideas into comprehensible, engaging narratives. In essence, we aim to fill the gap between information and understanding. At Mekiki Magazine, our readers do not just learn—they comprehend. They do not just read—they engage. And in doing so, they gain a richer, more nuanced view of the world around them.

Join us on a transformative journey of discovery with Mekiki Magazine, where cutting-edge technology merges with timeless wisdom to illuminate new horizons of knowledge and understanding.

www.ingramcontent.com/pod-product-compliance
Lightning Source LLC
Chambersburg PA
CBHW040759150726
48196CB00033B/870/J

9798892140096